STATISTICAL OPTIMIZATION OF BIOLOGICAL SYSTEMS

STATISTICAL OPTIMIZATION OF BIOLOGICAL SYSTEMS

TAPOBRATA PANDA
R. ARUN KUMAR
THOMAS THÉODORE

CRC Press
Taylor & Francis Group
Boca Raton London New York

CRC Press is an imprint of the
Taylor & Francis Group, an **informa** business

CRC Press
Taylor & Francis Group
6000 Broken Sound Parkway NW, Suite 300
Boca Raton, FL 33487-2742

First issued in paperback 2017

© 2016 by Taylor & Francis Group, LLC
CRC Press is an imprint of Taylor & Francis Group, an Informa business

No claim to original U.S. Government works

ISBN-13: 978-1-4665-8708-3 (hbk)
ISBN-13: 978-1-138-89313-9 (pbk)

Visit the Taylor & Francis Web site at
http://www.taylorandfrancis.com

and the CRC Press Web site at
http://www.crcpress.com

Tapobrata Panda dedicates this to his parents and teachers.
TP's inspiration for this work his wife, Meeta, and
daughters, Dr. Smriti and Miss. Shradha.

R. Arun Kumar dedicates to his father, Ravichandran; mother,
Manimegalai; brother, Rathna Kumar, and friends.

Thomas Théodore dedicates to his parents and sister.

Contents

Preface

All branches of basic and applied sciences deal with experimental approaches. In this age, enormous data are generated from the experiments. Sometimes, experiments and the approach to doing experiments are not properly guided by scientific methods. This consumes resources and does not lead to a meaningful understanding of systems. In keeping with this trend, experimenters use computational resources to solve their immediate needs without understanding the basic approach. The correct approach, however, is to properly design experiments by applying statistical knowledge. Mere use of statistical terms and their analysis, without knowledge of statistics, gives the user a black box concept. It is satisfying and fun to appropriately exploit statistics, which explains why, what, how, or when something happens, and to use a statistical experimental approach to obtain corresponding data, complete analysis, interpret results, or make predictions.

This book intends to engage the reader in developing a foundation for the statistical experimental approach. This is true for any branch of science or engineering.

The book addresses the problem of optimization of bioprocess systems in its entirety. There are a number of books written by statisticians who address the problem of mathematical optimization, but there is a void as far as the experimental optimization of biological systems employing statistical techniques is concerned. The optimization of biological systems requires a thorough understanding of the system and the application of the appropriate mathematical and statistical tools; this book fills this void.

The book is organized into seven chapters that emphasize the foundations and applications of biological systems.

Chapter 1 discusses the complex nature of biological systems and the need for optimization. The advantages and disadvantages of non-statistical methods of optimization and the rationale for statistical methods of optimization are presented. In Chapter 2, the traditional one-variable-at-a-time approach to optimization is reviewed with suitable examples. Chapter 3 deals with experimental designs. The planning of an experiment starts with the setting of goals or objectives and identifying the process variables. Then an appropriate screening design is chosen to identify the important variables, which have a profound effect on the outcome. The process must be made robust. Response surface methodology is then employed to maximize (or minimize) a response and to reduce variation by locating a region where the bioprocess is easier to manage. Then regression modelling is done to obtain a precise model, which would quantify the dependence of the response on the process variables. Chapter 4 details the statistical analysis and optimization of different experimental designs, the results of which are used to determine the important variables and optimum settings. The different optimization techniques employed to determine the optimum levels of the process variables for both single- and multiple-response systems are described. Chapters 5 and 6 outline less frequently employed experimental designs, like the Evolutionary Operation Programmes and Taguchi's designs, respectively. Chapter 7 describes the concept of hybrid experimental designs, which shows better optimization of data obtained from biological systems, using the essence of genetic algorithm. The entire book is illustrated with real-life bioprocess optimization problems and their solutions and exercise problems for practice. The book contains useful appendix, and an index.

Tapobrata Panda
Indian Institute of Technology Madras

R. Arun Kumar
Indian Institute of Technology Madras

Thomas Théodore
Siddaganga Institute of Technology
Tumkur

MATLAB® is a registered trademark of The MathWorks, Inc. For product information, please contact:

The MathWorks, Inc.
3 Apple Hill Drive
Natick, MA 01760-2098 USA
Tel: +1 508 647 7000
Fax: +1 508 647 7001
E-mail: info@mathworks.com
Web: www.mathworks.com

Acknowledgements

This book is the end product of three authors in various dimensions of biological experiments. About 20 years of research work on the statistical optimization of biological systems carried out under the supervision of Tapobrata Panda by different researchers in the Biochemical Engineering Laboratory of the Department of Chemical Engineering at the Indian Institute of Technology Madras is the basis of the book.

This is equally supplemented by the in-depth knowledge and practical experiences of Thomas Théodore for a decade in this field. However, extensive work of Arun Kumar in the fields of Taguchi's design and genetic algorithm-based hybrid design gives proper shape to this book.

We acknowledge the assistance of many people who have contributed in various ways to this project, particularly Dr. A. Seenivasan, assistant professor in the department of biotechnology, NIT Raipur, India, who cheerfully drew all the figures with great skill, and S. Venkatesan, IIT Madras, who typed the initial manuscript on time.

S. Ravi Kumar, Zeeshan, Digvijay, Bhargavi, Purab Jain, and Pooja were of immense help at various stages of the project.

We thank Dr. Gagandeep Singh and all the folks at Taylor & Francis Group for their understanding and unwavering support for this project and for flexible deadlines.

We thank our family members who put up with us working long hours on the book instead of spending time with them.

People around us credit us for this work. The responsibility for errors, oversight, and typos is entirely ours. We would greatly appreciate if you point them out, so that they may be corrected in a later edition.

Authors

Tapobrata Panda is currently a professor at IIT Madras. He was a visiting scientist in the Department of Chemical Engineering, Iowa State University, Ames, Iowa, and a visiting faculty at Asian Institute of Technology, Bangkok, Thailand. From the research team of Professor Panda at the IIT Madras, 22 PhD students have graduated by contributing in the areas of enzyme systems, kinetics, process optimization, and development of microbial products. His papers in peer-reviewed journals has 'h'-index (Google Scholar) of 30 and 'i-10' value of 64. He has authored a book *Bioreactors: Analysis and Design* (McGraw-Hill, 2011) for all people working in biotechnology. He has contributed chapters in the *Handbook Food Process Design* (Wiley-Blackwell) and in the *Encyclopedia of Metalloproteins* (Springer, 2013) in the avenue of gold nanoparticles biosynthesis. His current research areas are hybrid experimental design, Bio-MEMS, biological synthesis of nanoparticles, and the design of therapeutic molecules and enzyme design.

Professor Panda received his PhD in biochemical engineering from Indian Institute of Technology Delhi; MTech and BTech in food technology and biochemical engineering from Jadavpur University, Kolkata; and BSc (Honors) in chemistry from University of Calcutta, Kolkata. He has carried out advanced research at IBTM, Technical University of Vienna, Austria.

He was a member of the TAPPI, Peachtree Corners, Georgia. He is a member of the editorial board of the *Open Biotechnology Journal* (Bentham Science Publishers); the *Open Enzyme Inhibition Journal;* and *Advanced Science, Engineering, and Medicine Journal* (American Scientific Publishers).

R. Arun Kumar is currently working with an oil and gas super major in the liquefied natural gas business as a process engineer. Previously, he worked for an international oil and gas service company. He received his BTech in chemical engineering from the Indian Institute of Technology Madras. He was in the top 1% in the National Astronomy and Physics Olympiad. He worked in the areas of biochemical engineering, genetic algorithm applied to biological systems, and design of experiments.

Thomas Théodore is an associate professor of chemical engineering at the Siddaganga Institute of Technology, Tumkur, India. He obtained his PhD in biochemical engineering from the Indian Institute of Technology Madras. A chemical engineer from Annamalai University and AC College of Technology, Chennai, India, he also holds an MS degree in bioengineering from ESPCI Paris and a MEngSc in biopharmaceutical engineering from UCD Dublin. His areas of interest include therapeutic proteins and biodegradable polymers.

1

Introduction

OBJECTIVE: To provide overall knowledge and information required for understanding statistical experimental analysis of a biological system.

1.1 Why and How Biological Systems Differ from Their Counterparts?

A biological system is difficult to understand because of complex interaction among the biological components and the physical counterparts. This involves a series of process, operations, and interactions of living and non-living matter. A unit process may encounter with a single or multiple physical, chemical, or biochemical changes. Sometimes, they are combinations of changes. Rarely, a biological system has one single-unit operation. The dynamic nature of a biological system makes the situation more difficult to interpret in quantitative terms. To explain any system, it is better to quantify, which can translate the process in a more meaningful way. This quantitative information can be used to compare the efficiency of a process and, in turn, suggests the success of a process. If one tries to quantify a biological system, knowledge of various branches of science is necessary.[1] Table 1.1 tries to highlight the different branches required to explain and/or to quantify a chemical system and a biological system.

Hence, a biological system is more difficult to quantify and to interpret in terms of the variation of physical quantities. However, it is necessary to quantify the biological system at any cost, as tremendous opportunities exist for the present and future development in processes. If a biological system needs to be translated into a successful biological process, then it is necessary to know in details the factors involved in such a process.

1.2 Factors in Biological Systems

Identification of variables is important for a process to quality as well as to analyze for further developments of the process. The development of the process is associated with structured information of what experiments

TABLE 1.1

Knowledge Required Quantifying Chemical and Biological Systems

S. No.	Knowledge from the Branch	Chemical System	Biological System
1	Chemistry	Yes	Yes
2	Mathematics	Yes	Yes
3	Physics	Yes	Yes
4	Computer sciences	Yes	Yes
5	Biological sciences – Genetics – Biochemistry – Microbiology – Physiology	No	Yes
6	Medical sciences	No	Yes
7	Pharmaceutical sciences	No/limited	Yes
8	Environmental sciences	Yes	Yes

are required and need to be conducted in a meaningful manner. This suggests the implementation of non-statistically and statistically established experimental approaches. In this regard, generalization of unique factors across the biological systems (or in precise the biological process) is not possible. Therefore, this section will highlight factors for a few of the biological processes.

1.2.1 Fermentation Process

This process is a combination of upstream operations, on-going process, and the downstream operations. Let us consider them separately.

1.2.1.1 Upstream Operations

Varieties of upstream operations are sterilization of air required for aerobic fermentation, formulation of medium for reactions, thermal sterilization of medium constituents, sterilization of reactor for biological reactions, and processing of raw materials for biological reactions. Table 1.2 presents the factors to understand the system for experimentation.

1.2.1.2 During Fermentation

The factors are controlled pH of the reacting fluid, temperature of fermentation, agitation speed, aeration rate (for aerobic fermentation), feed-flow rate (for flow bioreactor operation), pCO_2, antifoam dosage (for chemical antifoam system), and periodicity of feeding of reactants (for fed-batch and batch-fed operations). A few examples in Table 1.2 give the concept of variables influencing the process.

TABLE 1.2

Factors in Some Selected Operations in Fermentation Process

Biological Process Operations	Factors
1. Upstream Operations a. Sterilization of air	*For fibre filter*: Dimension of filter, packing density of fibre, packing length, pressure drop, airflow rate, input microbial load, and type of organism in incoming air to filter *For membrane filter*: Type of membrane, membrane dimension, pore size in the membrane, allowable pressure, and fouling components
b. Sterilization of medium constituents	*For batch operation*: Composition of medium components and time of exposure *For continuous operation*: Flow rate of medium components, physical properties of solution, length of holding section of the sterilizer, and contaminant load
c. Processing of medium components	*For bio-solubilization of coal*: Particle size and conditions of pre-treatments *For the hydrolysis of agricultural residues*: (1) size of particles, (2) types of acid/alkali, (3) amount of acid/alkali used, (4) strength of the acid, (5) contact time of acid/alkali with agricultural residues, and (6) mixing condition *Using steam*: (1) steam pressure, (2) saturation/super saturation, and (3) contact time *For the hydrolysis of cellulose/hemicellulose*: (1) particle size, (2) conditions of acid/alkali used as indicated in hydrolysis of agricultural residues, (3) pH of the hydrolysis condition, (4) temperature for hydrolysis, and (5) mixing condition
d. Reaction medium formulation	*For extracellular enzyme production, namely, chitinase*: KH_2PO_4, NaH_2PO_4, $MgSO_4$, $ZnSO_4$, and $CaCl_2$ *Other associated variables*: The pH of reaction medium, temperature, age of inoculums, and amount of inoculums
2. Conversion of reactant	The pH of reaction medium, temperature reactant concentration, volume of enzyme (may be expressed in terms of protein content), incubation time, and stirring condition of reaction medium
3. Synthesis of fused cells	Conversion of protoplast and sphaeroplasts to fusants and heterokaryons, pulse voltage (volt/mm), pulse width (µs), and number of pulses
4. Fermentation process	*Production of an organic acid*: The pH of reaction medium, aeration, agitation, and time of fermentation
5. Assay of an enzyme	*Cells associated enzyme extraction*: Initial activity of enzyme, time of incubation, temperature of assay, and the pH of reaction condition
6. Enzyme production	*Medium composition*: Glucose, peptone, yeast extract, malt extract, speed of agitation, pH of the medium, temperature, slant age, inoculums age, and cell concentration
7. Extraction of an intracellular product	*Polymer*: Temperature, pH of extraction fluid, digestion time, hypochlorite concentration, and sodium dodecylsulphate concentration
8. Hydrolysis of polymeric substances	*Pectin by polymethylgalacturonase*: Volume of substrate, volume of enzyme, pH, and temperature
9. Fermentation in aerobic condition	pH, aeration rate, and agitator speed

1.2.1.3 Downstream Operations

Product isolation, concentration, and purification from fermentation broth have different factors, which required optimization. For example, the isolation of a product from cells is affected by the extractant volume, contact time of extractant with cells, agitation speed during extraction, temperature, cycles of operation, and so on. Purification of a product in a chromatographic system needs careful consideration of element composition, flow rate of element, packing nature of chromatographic system, temperature, pH of element (in some cases), and gradient cycle time (in specific cases). Table 1.2 contains some relevant examples.

1.2.1.4 Special Biological Systems

For the synthesis of hybrid cells, characterization of products obtained from biological reactions and analysis of products are some of the special biological systems, which have different set of factors (*cf.* Table 1.2 for specific examples). Readers may get more detailed nature of factors for the given biological systems from other chapters of this book. The experiments need to carefully observe and analyze such factors involved in the processes.

1.2.2 Classification of Factors

In Table 1.2, a few factors have been indicated for a given process. This is not exhaustive as the process may be affected by other factors whose effect may be less than the factors mentioned for a particular process. For example, the fermentation using fungi is affected by medium components, the pH of reacting fluid, temperature, cell mass density, and age of cells. However, other less important factors are mycelia growth rate, number of hyphal tip, conidial development and so on.

In a general sense, one may classify the factors involved in biological systems as follows:

- Chemical factors (e.g. medium constituents)
- Physical factors (e.g. temperature, shear force, and agitation rate)
- Physicochemical factors (e.g. the pH of reacting fluid and redox)

In a broader sense, following the concept of Montgomery,[2] the factors in biological experimentation can be classified as per Figure 1.1.

Potential factors are typically considered for an experimental plan. For example, in medium optimization, one can find 17 medium constituents strongly influencing the production. They are considered as design factors and rests of the factors are apparent constant factors. The process is carried out in a confined system, for example, a bioreactor for production. If one uses same volume of bioreactor of different configuration, then production is certainly influenced by such variation, which is termed as *allowable factors*.

Potential Factors with Respect to Experimental Plan			Unrelated Factors According to Experimental Plan		
Design factors	Apparent constant factors	Allowable factors	Controllable	Uncontrollable	Noise

FIGURE 1.1
Classification of factors.

To carry out the experimental plan for a biological system, the effects of *allowable factors* have been assumed to be very small.

The role of related factors may be significant with respect to a particular process. However, they have not given importance, as a particular experimental plan does not consider their role. Sometimes, the levels of *unrelated factors* can be set to a particular value, which is termed as *controllable unrelated factors*. If the *unrelated factors* are difficult to estimate, which eventually are difficult to control, they are called *uncontrollable unrelated factors*. Natural and uncontrollable variation is called the *noise factor*. One can adjust the controlled design factor to minimize the variation due to *noise factors*. The formulation of a biological process thus involves various factors in isolation or in a combination thereof. To get comprehensive knowledge of the process, a clear understanding of factors is possible by optimization of the involved components.

Terminologies are defined here to have better understanding of this book.

1.3 Terminologies

1.3.1 Replication

Independent repetition of factor combination is called the *replication of the experiment*. Independent fermentation has been carried out at the pH of reaction medium 5 and at temperature of fermentation of 25°C using *Aspergillus niger*. In this case, the number of replication is three.

1.3.1.1 Reasons for Replication

The following are the reasons for replication:

1. It facilitates estimation of experimental error.
2. Precise estimation of the parameter is possible.
3. This is a means to know about the variation. This can allow the estimation of variance (σ^2) of an individual observation and variance of sample mean (σ_0^2). For N number of replicates,

$$\sigma_0^2 = \frac{\sigma^2}{N} \tag{1.1}$$

4. This provides sources of variation between runs and within runs.

1.3.2 Repetition of Assay

In the above example given in replication, one can estimate a parameter, that is, dry cell weight of *A. niger*. From the same experiments, if four samples are estimated for dry cell weight, then it cannot be considered as replication. This is repetition.

1.3.3 Random Effect

If experiments are carried out in a random fashion, external effects may be evenly distributed in the experimental procedure. This is one of the requirements in statistical methods. Commercial software for experimental design generates run order number and experiment member. Complete randomization is an ideal procedure, which is costlier as well.

1.3.4 Blocking

This is a procedure to improve precision for experimentation. Blocking reduces the variation caused by the unrelated factors according to experimental plan.

1.3.5 Levels and Range of Factors

The term *level* is used to describe a particular property of the factor. The high and low pH values of biological reactions are considered as levels of factors.

1.3.6 Errors

In experiments, one needs to get correct and reliable data. Standard procedures are followed in this regard. However, no measurement can be considered as absolute. There will be some deviation from the ideal value, called an *error*. It can be associated with the experiment followed and the measurement made in all stages of the experiments. For insignificant errors, there is no problem in assigning the conclusion from the experiment. If the error is significant, one cannot conclude from the experiments.

Due to faulty concept of the experiments, the experimental results will not converge. This gives an idea of *gross error*,[3] which is not discussed further for statistically designed experimentation.

The measurement technique has some obvious error associated with it. For example, the estimation of lovastatin by spectrophotometer always gives lower result than the high performance liquid chromatography (HPLC) technique.[4] Some experiments follow the method using spectrophotometer with a prior knowledge of the technique to estimate lovastatin, called a *systematic error*. This can be minimized using sophisticated analytical tool, for example, HPLC.

If one estimate lovastatin by the HPLC technique for the same source of sample *n* number of times, then the results do not exactly match. This type of error influences the outcome of the experiment. The experimenter does not exactly know the prediction of the variability on the result. This is called a *random error*.

1.3.6.1 Types of Errors in Statistically Designed Experimentation

Two different types of errors[3] are considered for discussion in the measurement in this book.

Type 1 error

This rejects the test hypothesis when it is correct.

Reason

The significant level of a test is the probability that the observed value of test statistic will be in the rejection range.

Frank and Altheon[5] defined Type 1 and 2 errors.

$$\alpha = p\left(H_0 \text{ rejected, when } H_0 \text{ is correct}\right) \tag{1.2}$$

where:

α is the choice of significance level

p is the probability

H_0 is the null hypothesis

Type 2 error

This rejects the alternative hypothesis (H_{alt}) when it is correct. This is denoted by β.

$$\beta = p\left(H_{alt} \text{ rejected, when } H_{alt} \text{ is correct}\right) \tag{1.3}$$

α and β are inversely related.[5]

The experimenter needs to specify a particular distribution for the test statistic to evaluate β. Table 1.3 tabulates some differences between Type 1 and Type 2 errors.

1.3.7 Sample

A collection of finite or infinite individual value is called *population*. A sample is a part of population, having a proper objective to make a conclusion

TABLE 1.3

Difference in Type 1 and 2 Errors

S. No.	Type 1 Error	Type 2 Error
1	Null hypothesis (H_0) is rejected when it is correct	H_0 is accepted when it is false; H_{alt} is rejected when H_{alt} is correct
2	It leads to false positive results	It leads to false negative results
3	Sample and size depend on the probability of choice of significance level	Small, that is, depends on distribution of test statistic

about the population. We classify them as representative and random samples.

In representative samples, relevant characteristics are present as in the population. On the other hand, a set of value is drawn in a manner that individual set has the same chance to be present in the population.

1.3.8 Sample Size

The experimental design considers how one will collect data and what will be the number of data. With the increase in sample size, variability decreases.

1.3.8.1 Parameters to Evaluate

A few parameters are necessary to be defined here.

Sample average: It is defined by Equation 1.4.

$$\bar{X} = \frac{1}{N} \sum_{u=1}^{N} x_u \tag{1.4}$$

where:
  x_u is the u_{th} value
  N is the sample number
  $\bar{X}$ has the same unit of x_u

Sample variance: Equation 1.5 describes the sample (or population) variance.

$$\text{var}(x) = \frac{1}{N-1} \sum_{u=1}^{N} (x_u - \bar{x})^2 \tag{1.5}$$

where:
  $(x_u - \bar{x})$ = deviation of x_u about sample average
  $\text{var}(x)$ has the unit of square of the unit of x_u

Sample standard deviation (sd): sd is the $\sqrt{\text{var}(x)}$ having unit of x_u. In other words,

$$(\text{sd})^2 = \text{var}(x)$$

where:

sd is the intervals about the sample average

1.3.9 Variables

In biological processes, the output of experimentation is called the *response variable* (Y). Y depends on various factors of experiments. Y is called the *dependent variable* and the factors properly coded are called the *independent variables*. In experiments, variables are classified into continuous and categorical. Rates, proportion, and percentage used to obtain information are called the *categorical variables*. Data pertaining to categorical variables are classified into ordered categories and non-ordered categories. This is described in detail by Peat and Barton.[6]

1.3.9.1 Continuous Variables

Data involving continuous variables are dealt here in the following sections.

- Distribution of the variables
- Comparison of two independent samples
- Analysis of variance (ANOVA) and analysis of covariance (ANCOVA)
- Correlation and regression

1.3.10 Covariance

The mean of the products of deviation of two variables (x_1 and x_2), for example, is the co-variance (Covar) of x_1 and x_2.

$$\text{Covar}_{x_1,x_2} = \frac{\sum\limits_{i}^{N}(x_{1_i} - \bar{x}_1)(x_{2_i} - \bar{x}_2)}{N} \tag{1.6}$$

where:

N is the number of observations

1.3.10.1 Properties of Covariance

- For statistically unrelated variables, $\text{Covar}_{x_1,x_2} = 0$.
- The sign of Covar is an indication of either positive or negative correlation between the variables.
- The magnitude of Covar suggests how strongly or weakly the variables are correlated between them.

1.3.11 Variance

The mean squared deviation from the mean is called the *variance*. From absolute deviation, it is expressed as per Equation 1.7.

$$\text{var}(x) = \frac{\sum_{}^{n}(x_u - \bar{x})}{n} \tag{1.7}$$

Frank and Althoen[5] defined variance as per Equation 1.8.

$$\text{var}(x) = \frac{\sum_{}^{n'}(x_u - \bar{x})f_i}{n} \tag{1.8}$$

where:
  f_i is the frequency

1.3.11.1 *How Shall One Calculate Variance?*

Variance can be calculated using Table 1.4.

$$\bar{x} = \frac{\sum x_u}{\text{Number of observations}}$$

Then one can use Equation 1.7 to calculate var(x).

1.3.11.1.1 *Calculation of Variance Using Frequency Data*

The technique is described in Table 1.5.
Therefore, var(x) can be calculated using Equation 1.8.
  Frank and Altheon[5] have also defined variance and sd using Equations 1.9 and 1.10.

TABLE 1.4

Calculation of Variance

Number of Observation	x_u	$(x_u - \bar{x})$	$(x_u - \bar{x})^2$
1	x_1	$(x_1 - \bar{x})$	$(x_1 - \bar{x})^2$
2	x_2	$(x_2 - \bar{x})$	$(x_2 - \bar{x})^2$
.....	.....	.....	.....
i	x_i	$(x_i - \bar{x})$	$(x_i - \bar{x})^2$

TABLE 1.5

Use of Frequency Data for var(x) Calculation

Number of Observation	x_u	f	$x_u f$	$(x_u - \bar{x})$	$(x_u - \bar{x})^2$	$(x_u - \bar{x})^2 f$
1	x_{u_1}	f_1	$x_{u_1} f_1$	$(x_{u_1} - \bar{x})$	$(x_{u_1} - \bar{x})^2$	$(x_{u_1} - \bar{x})^2 f_1$
2	x_{u_2}	f_2	$x_{u_2} f_2$	$(x_{u_2} - \bar{x})$	$(x_{u_2} - \bar{x})^2$	$(x_{u_2} - \bar{x})^2 f_2$
.....	.....	.....	.....	.....	.....	.....
i	x_{u_i}	f_i	$x_{u_i} f_i$	$(x_{u_i} - \bar{x})$	$(x_{u_i} - \bar{x})^2$	$(x_{u_i} - \bar{x})^2 f_i$
		Σf	$\Sigma x_u f$			

$$\bar{x} = \frac{\Sigma x_u f}{\Sigma f}$$

$$\text{var}(x) = \frac{\sum^{n'} x_u^2}{n} - \bar{x}^2 = \frac{\sum^{n'} x_u^2 f_i^2}{n} - \bar{x}^2$$

$$\text{and} \tag{1.9}$$

$$\text{sd} = \sqrt{\frac{\sum^{n'} x_u^2}{n} - \bar{x}^2}$$

$$= \sqrt{\frac{\sum^{n'} x_u^2 f_i^2}{n} - \bar{x}^2} \tag{1.10}$$

1.3.12 Coefficient of Variation

It is defined as per Equations 1.11 and 1.12.

$$\text{Coefficient of variation} = \left(\frac{\text{sd}}{\bar{x}}\right) \times 100 \tag{1.11}$$

This is useful for the following reasons:

- Comparison of data set having different means
- Comparison of data set having different units

This measures the unexplained or the variation due to residual as a percentage of mean of the response variable.

$$\text{Coefficient of variation} = \left(\frac{\sqrt{\text{MSE}}}{\hat{Y}}\right) \times 100 \tag{1.12}$$

where:
   MSE is the mean standard error

1.3.13 Standard Error

For n independent observation, if the variance is defined by var(x), the standard error (SE) is expressed by Equation 1.13.

$$SE = \sqrt{\frac{var(x)}{n}} = \frac{sd}{\sqrt{n}} \tag{1.13}$$

This is also called the *SE of the mean.*

  Significance: If n increases, variation in the sample mean will decrease.

1.3.13.1 Differences between sd and SE

Table 1.6 summarizes the difference between sd and SE.

1.3.14 Correlation Coefficient

According to Frank and Althoen,[5] the correlation coefficient between two variables is defined by $r_{x_1x_2}$ by the following formula (Equation 1.14):

$$r_{x_1x_2} = \frac{Covar_{x_1x_2}}{\left(sd_{x_1}\right)\left(sd_{x_2}\right)} \tag{1.14}$$

Properties of correlation coefficient: If $r_{x_1x_2} = \pm 1$, there is a perfect correlation between the variable (e.g. x_1 and x_2). The relation between x_1 and x_2 is linear.

1.3.15 Correlation Ratio

The relation is as follows:

$$\text{Correlation ratio} = \frac{\left(\begin{array}{l}\text{Variance of marginal}\\\text{distribution}\end{array}\right)_Y - \left(\begin{array}{l}\text{Conditional}\\\text{variance}\end{array}\right)_{Y/x}}{\left(\text{Variance of marginal distribution}\right)_Y}$$

TABLE 1.6

Major Points in sd and SE

Standard Deviation	Standard Error
Quantifies scatter, varies from sample to sample	Quantifies precisely; one gets true mean of the population
Same on average for increased sample size	Considers both sd and sample size
Units of the data	Units of the data
sd is almost unchanged with larger sample size	SE gets smaller with larger sample size
sd does not describe the accuracy of the sample mean	SE does not describe the variation of individual values

1.3.16 Coefficient of Determination

For linear relationship between the variables, correlation ratio is the square of correlation coefficient for variables ($r^2_{x_1 x_2}$ or R^2). This is called the *coefficient of determination*.

If x_1 and x_2 are not linearly related, the coefficient of determination is less than the correlation ratio. R^2 is a measure of the percent variance in the outcome variable. The following relation expresses R^2.

$$R^2 = \frac{\text{Sum of squares residuals}}{\text{Sum of square total}} \quad (1.15)$$

This statistic suggests the measure of total variation of the observed values around the mean value obtained by the fitted model.

1.3.17 Adjusted R^2

This is a variation of the R^2 statistic. A number of factors in the model can be obtained from this statistic.

1.3.18 R^2 for Prediction

This is based on calculating R^2 in terms of prediction error sums squares.

1.3.19 Adequate Precision

This is defined by Equation 1.16.

$$\text{Adequate precision} = \frac{Y_{\text{maxpred.}} - Y_{\text{minpred.}}}{\text{sd}_{\text{average of all predicted responses}}} \quad (1.16)$$

1.3.20 Confidence Interval

This is a range defined by lower limit (LL) and upper limit (UL), which preferably a true parameter value would lie in this interval, with proper degree of confidence.

$$\text{Confidence interval} = \text{mean} \pm (1.96 \times \text{SE}) \quad (1.17)$$

Let one assume τ, an unknown parameter. To estimate the interval for τ, LL and UL are calculated with a given probability condition:

$$p(\text{LL} \leq \tau \leq \text{UL}) = (1 - \alpha) \quad (1.18)$$

is true.

Then, as per Montgomery,[2] the interval $\text{LL} \leq \tau \leq \text{UL}$ is $(1 - \alpha)$ 100% confidence interval for τ. $(1 - \alpha)$ is called the *confidence coefficient*[2]. Suppose $\alpha = 0.01$, then, $\text{LL} \leq \tau \leq \text{UL}$ is called a 99% confidence interval for τ.

Confidence intervals give an estimate on the precision around a statistic. The statement like 90% confidence interval means the statement or a statistic will be true in 90% of occasions. In most of the analysis, 95% confidence intervals are stated in the discussion. One can have a 95% confidence interval from the following example.

$$t_{(n-p)}(0.025) \leq \frac{(\text{Projection coefficient}) - (\text{Expected value})}{\text{SE}} \leq t_{(n-p)}(0.975)$$

where:
$(n-p)$ is the degrees of freedom

1.3.21 Outliers

Data points lying more than three sd from the regression are called *outliers*. There are two types of outliers: univariate and multivariate.[6]

Univariate outliers: If the absolute z-score is greater than 3, the outliers are univariate outliers. For small sample size, z-score is more than 2.5.

Multivariate outliers: If the data points have extreme values as a combination of explanatory variable, then they are called *multivariate outliers*.

Outliers cause worry to the experiments, which might invalidate the analysis of the experiments. To correct the outlier, one can consider an estimate to replace the outlier. A detailed treatment of outlier is described by Montgomery.[2]

1.3.22 ANOVA

ANOVA is a broad terminology for standard statistical techniques. This is not only group techniques but also a method of constructing statistical model for experimental component. The model can be expressed by the approach of Johnson and Leone.[7]

$$Y_{\text{obs}} = \sum (\text{Parameters represent assigned effects})$$
$$+ \sum (\text{Random variables for assigned effects}) \quad (1.19)$$
$$+ \sum (\text{Random variables for unaccounted effects})$$

where:
Y_{obs} is the observed value

In biological reactions, concentration of reactants, temperature, and pH are assigned effects. Unaccounted effects are batch variation, cell behaviour and so on.

Johnson and Leone[7] suggested the following four assumptions about the random variable and parameters:

- All variation in expected value of residual random variable is covered by parameters representing assigned effects.
- Mutual independence for residual random variable.
- Normal distribution for each random variable.
- Same standard deviation for residual random variables.

1.3.22.1 ANOVA Table

The calculation of a test statistic, namely *F*-statistic, and the orthogonal break up of squared length of an observed vector $\|Y\|^2$ using Pythagoras theorem is summarized in Table 1.7, called the *ANOVA table*.

One can get one-way and two-way classifications for one factor and two or more factors, respectively. *F*-statistic is defined by Saville and Wood.[8]

$$F = \frac{\|Y\|^2}{\left(\left\|\frac{Y - \bar{Y}}{2}\right\|^2\right)} \qquad (1.20)$$

In one-way classification, the calculation of SS are *between groups* and residual (*within groups*).

The *two-way classification* is divided into *hierarchal* and *cross* classification. In *hierarchal* classification, the calculation of SS involves between main groups, between subgroups within main groups, and residual (within subgroups).

On the other hand, cross-classification involves SS calculation between rows, between columns, and residual (within *cells*).

We shall indicate the ANOVA table for balanced – incomplete – block design and factorial designs.

1.3.22.1.1 Multivariable Analysis of Variance

Univariate analysis of variance explained above for a single response can be extended for multi-response analysis. For example, in a biological system, cell growth and a specific product formation are considered as two responses. The

TABLE 1.7

Classical ANOVA Table

Type of Variation	Degrees of Freedom (df)	Sum Squares (SS)	Mean Square (MS)	*F*-Statistic
Subspace	Dimension of Subspace	SS of Projection Length	Mean Squared Length	Ratio of Mean Squares
Mean				
Error				
Total				

resultant technique for multi-responses can be applied to test the hypotheses on parameters in the model for each of the variables in question.

1.3.22.1.2 *Steps of ANOVA*

The steps involved in ANOVA are as follows:

- Define an objective of the experiment
- Design the experiment
- Data obtained from the experiment
- Proposing a model to understand data – this includes assumptions
- Test hypothesis: Testing the null hypothesis $H_0 : \beta_1 = 0$ against the alternative $H_{alt} : \beta_1 \neq 0$, where β_1 is the slope of unknown true line (Figure 1.2)
- To fit the said model
- To test the hypothesis
- To formulate the ANOVA table
- To estimate of variance (var)
- To analyze data in the ANOVA table
- To check the assumptions
- To refit the model with the introduction of revised ANOVA table
- Transform the curve
- To prepare a report on the study

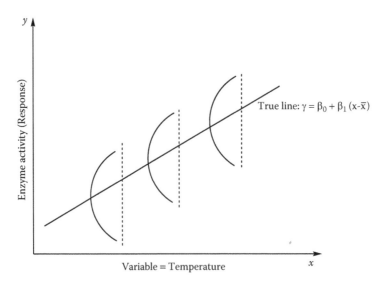

FIGURE 1.2
Relation of response variable and input variable.

1.3.23 Commonly Used Statistics

1.3.23.1 t-Statistic

This is defined by Equation (1.21).

$$t_{df} = \frac{V_1}{\sqrt{\left(V_2^2 + \cdots + V_{(df+1)}^2\right)/df}} \tag{1.21}$$

where:
$V_1 \ldots V_{(df+1)}$ are independent random variables
t_{df} can have positive and negative values

The numerator has a normal distribution. Therefore, t_{df} is symmetric about zero. The denominator is the square root of variance. If the degree of freedom increases, the variability of this estimator decreases to zero.

Therefore, the t-distribution is considered as a special case of F-distribution for $p = 1$[8]. They have defined t-statistic by Equation 1.22.

$$t = \frac{\text{Estimate of the parameter}}{\text{Standard error of the estimator}} \tag{1.22}$$

1.3.23.2 F-Statistic

F-statistic is the ratios of averages of squared projection lengths. This follows a distribution unlike t-distribution.

The F-statistic is expressed by Equation 1.23.

$$F_{m,n} = \frac{\left[\left(V_1^2 + \cdots + V_m^2\right)/m\right]}{\left[\left(V_{m+1}^2 + \cdots + V_{m+n}^2\right)/n\right]} \tag{1.23}$$

where:
$V_1, \ldots, V_{m+n}$ are independent random variables
m and n are degrees of freedom

1.3.23.3 Chi-Square Statistic

This statistic is used to measure a condition that is significantly different between two or more groups.

$$\text{Chi-square statistic} = \frac{\sum \left(\text{Observed value} - \text{Expected value}\right)^2}{\text{Expected value}} \tag{1.24}$$

1.3.24 ANCOVA

ANCOVA is a mixture of techniques of ANOVA and regression.[6]

1.3.24.1 Application

It is used to analyze group differences after adjusting the outcome variable for a continuously distributed covariate.

The assumptions of ANCOVA and ANOVA are the same; however, ANCOVA has the following additional assumptions:

- Reliability in the measurement of covariate
- Low colinearity in covariates for multiple covariates
- Linear relation between covariate and outcome
- Homogenous regression
- No interaction between covariate and factors

1.3.25 Testing of Hypothesis

Hypothesis about an experiment is a belief or doubt, which is provisionally considered. In this regard, one needs the proof to establish the hypothesis or to reject the hypothesis. The steps of testing hypothesis are as follows:

- Formulation
- Decisions
- Data collection
- Conclusion

A detailed theory is described by Frank and Althoen.[5]

1.3.26 Data Analysis

The model equation (M_1) can be represented by Equation 1.25.

$$\text{Data} = \text{Model} + \text{Error} \tag{1.25}$$

$$Y_i = \beta_0 + \text{Error} \tag{1.26}$$

One can say β_0 is unknown. One can estimate it. To make conditional predictions, β_0 adjusted by a quantity (suppose that β_1) is to reduce the error.
Therefore,

$$Y_i = \beta_0 + \beta_1 + \text{Error subjected to conditions} \tag{1.27}$$

Assuming that x_i is the amount of states proportion, Equation 1.27 becomes Equation 1.28.

$$Y_i = \beta_o + \beta_1 x_i + \text{Error} \qquad (1.28)$$

We try another model M_2 to replace M_1. M_2 contains all parameters of earlier model plus additional parameters. The additional parameters (1) may or (2) may not reduce errors.

Suppose error (in using model M_2) $\leq$ error (in using model M_1). One can calculate proportional reduction in error. Then, one can decide on the role of additional parameters in the models. Judd et al. [9] and Fisher[10] have given detailed techniques for data analysis.

1.4 What Is Optimization?

In a biological system, one can see, from Section 1.2, that there are large numbers of variables. However, one has to decide on solving the problems. Optimization is the centre of making such decision. The setting of input variables is made to obtain maximum response (Y) values. Those input variables are now called *optimum conditions* for a particular output of a biological unit process.

The steps are as follows:

1. Select a process whose optimum conditions need identification.
2. List the input variables affecting the biological unit system.
3. Identify the input variables having greatest influence on the response values (Y).
4. Postulate a model in which the response values (Y) is the function of the input variables. If sufficient information is not available to establish the relation between Y and the input variables, first-order polynomial equation can be considered as the starting point of postulation, that is, a relation between Y and the input variables.
5. Carry out the initial set of experiments to collect sufficient information about the role of input variables on Y.
6. Now, refine the postulate stated in step 4, which is economical in experimentation.

We shall deal with detailed optimal condition in subsequent Chapters 2 through 7. Basics of optimization are dealt in details by Chong and Żak.[11] Figure 1.3 provides an overview.

Optimal settings of input variables can be achieved by the following procedures:

- Non-statistical experimentation
- Statistical experimentation

OPTIMIZATION

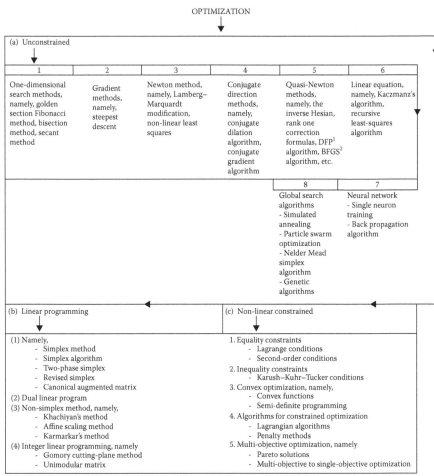

FIGURE 1.3
Various optimization techniques.

Non-statistical experiments are either the technique of one-factor-at-a-time from the pool of responsible variables or the movement of simplex techniques. Non-statistical experiments are useful when no detailed background information is available for the initial value of the variables and/or their experimental range. For example, the experimenter has isolated a strain of *Aspergillus* sp. Actual growth and cultivation conditions are not known for this strain. Hence, a few initial trial experiments are carried out to understand the type and number of variables. The organism is first planned to grow in the Czapek-Dox medium. The chemical composition of the Czapek-Dox medium is not optimal composition. The organism grows on different levels of medium composition. Not all variables (e.g. chemical factors) can be varied at the same time. One variable at different levels are studied, keeping other variables at fixed level. In this way, the particular variable at one concentration gives higher product concentration (in this case, dry cell weight equivalent). This particular concentration is called the most suitable concentration of the variable under examination. The levels (settings) of the variables obtained in this manner do not give the optimal levels of the variables. This is called *more suitable level of variables*.

1.4.1 Application

These experiments give an idea for the range and type of the variables, which may be considered for obtaining optimal values. In the scope of this book, the statistical experimental plans are described in Chapters 3 and 4 to achieve the optimal settings of the variables.

1.4.1.1 Solved Example

Based on different parameters defined above, following example is a practice to understand the equations provided in this chapter.

The experiments have been carried out in a controlled batch bioreactor. Citric acid concentration has been measured in the sample during fermentation. The fermentation has been carried out for six days. The product synthesized represents the information pertaining to the seventh day of fermentation in Table 1.8. One has to calculate the following:

1. Sample size
2. Sample average
3. Sample variance
4. Standard deviation

Solution:

1. Sample size is 16.
2. Sample average: From Equation 1.4, $\bar{X} = 0.4325$
3. Sample variance: From Equation 1.5, $\text{var}(x) = 0.0058$
4. Standard deviation is 0.0759

TABLE 1.8

Citric Acid Biosynthesis in Batch Experiments

Batch Experiment Number	Product Synthesized by Cell (kg/m³)
1	0.4816
2	0.3750
3	0.4823
4	0.5105
5	0.4203
6	0.3647
7	0.3933
8	0.4859
9	0.6515
10	0.4420
11	0.4218
12	0.3858
13	0.3840
14	0.3803
15	0.3657
16	0.3760

Exercises

1.1 Chitinase synthesis has been carried out in a separate batch experiment using a fungus. Thirteen planned experiments have been carried out. Mean product synthesis is 0.3494 units. Sum square total is 0.0248 with 12 degrees of freedom. The sum square regression is 0.0245 with 5 degrees of freedom. Calculate sum square error, mean square regression, and mean square residual.

1.2 Obtain the information from Problem 1.1. Construct the ANOVA table.

1.3 Establish the relation between coefficient of determination (= R^2) and F-statistic.

1.4 Calculate confidence level for the experimental runs to identify important variables to design growth medium for *Escherichia coli*.

Variable	t for H_0 Parameter $= 0$	Probability
A	2.98	0.0309
B	0.753	0.4855
C	−1.582	0.1744

References

1. Panda T, Babu PSR, Kumari JA, Rao DS, Théodore K, Rao KJ, Sivakesava S et al., Bioprocess optimization – A challenge, *Journal of Microbiology and Biotechnology*, 7(6), 367–372, 1997.
2. Montgomery DC (Ed.), *Design and Analysis of Experiments*, 7th edition., Wiley, New Delhi, India, 2009.
3. Bruns RE, Scarminio IS, and de Barros Neto B (Eds.), *Statistical Design – Chemometrics*, Elsevier, Amsterdam, the Netherlands, 2006.
4. Seenivasan A, Studies on biosynthesis of lovastatin and its analogues, PhD Thesis, Indian Institute of Technology, Madras, India, 2013.
5. Frank H and Althoen SC (Eds.), *Statistics – Concepts and Applications*, Cambridge University Press, Cambridge, UK, 1994.
6. Peat J and Barton B (Eds.), *Medical Statistics – A Guide to Data Analysis and Critical Appraisal*, Blackwell, Oxford, 2005.
7. Johnson NL and Leone FC (Eds.), *Statistics and Experimental Design in Engineering and Physical Sciences,* Vol. II, John Wiley & Sons, New York, 1977.
8. Saville DJ and Wood GR (Eds.), *Statistical Methods: The Geometric Approach*, Springer-Verlag, New York, 1991.
9. Judd CM, Mc Cleland GH, and Ryan CS (Eds.), *Data analysis: A Model Comparison Approach*, 2nd edition, Taylor & Francis Group, Routledge, New York, 2009.
10. Fisher NI (Ed.), *Statistical Analysis of Circular Data*, Cambridge University Press, Cambridge, UK, 1993.
11. Chong EKP and Żak SH (Eds.), *An Introduction to Optimization*, 4th edition, Wiley, Hoboken, NJ, 2013.

2

Non-Statistical Experimental Design

OBJECTIVE: The basic concept of an experimental approach is the guideline for traditional technique to find a suitable condition for the variable. In the self-directing experimental design, the response obtained from the experiment is the guideline.

2.1 Introduction

2.1.1 What Is an Experiment?

In a simple fashion, an experiment is a trial or a series of trials that suggest necessary alteration for input variable of a biological system for the identification of the roles of input variables to the output response. The overall objective of an experiment is to develop the technology. Experiments are random, sequential, or designed. In some cases, the demand is for the development of robust technology. Mode of experiments suggests different experimental designs (Figure 2.1).

All experimental procedures may not involve statistics in depth, namely, self-directing experimental design (SDED) and one factor at a time (OFAT)/ one variable at a time (OVAT) at all or very basic statistics, namely, evolutionary operation programme, some influences of statistics, the robust design of Taguchi, and complete statistical analysis, namely, response surface designs with single-response and multi-response strategies. OFAT/OVAT sometimes requires response surface models of first- and/or second-order models.[1] This will be discussed in Chapter 3. However, Box[2] suggested the following general criteria for *good experimental designs*:

- To provide satisfactory distribution of information
- To give the model value closest to the true value
- To allow transformation
- To have the activity to test for lack of fit
- To allow blocking experiments
- To allow sequential assembly
- To get an estimate of internal error

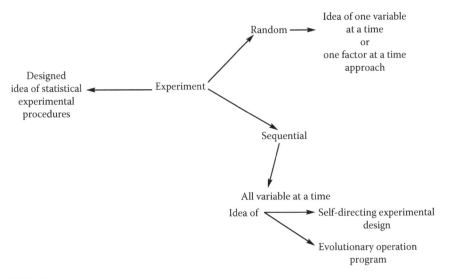

FIGURE 2.1
Strategies of Experimentation.

- To show sensitivity to outliers
- To have minimum number of runs
- To allow a visual understanding of information in the data
- To have a simple calculation
- To have the capability to respond to errors in variables
- To consider only a few levels of variables
- To provide a check of the constant variance and assumption

Some of the stated criteria do not fit well when the experimental design considers information that is absolutely required to explain the strategy of experimentation.

2.2 Steps in Designing an Experiment

A plan to do an experiment is associated with specific objective, namely, selection of preference for choice of variables (e.g. screening design), confirmation of the results/response, optimization of variables to run a system or process effectively, sensitivity and stability of the system (i.e. how far is it stable with the variation of input variables?), and discovery and further invention. It has three important subsections:

1. Statement of the problem
2. Selection of response, yield, or targeted result, considering single- and/or multi-response concepts
3. Selection of input conceivable variables, following refining of variables into specific measurable forms, segregation of variables (e.g. physical, chemical, physiochemical, and biological), and coupling of variables

   - Selection of targeted variables with their individual ranges applicable for experiment
   - Choice of experimental approach as per Figure 2.1
   - Methodology adopted and their standardization for the measurement of input and output variables
   - To conduct experiment with sufficient replicates and accuracy
   - Methods such as, statistical, non-statistical, kinetics and so on are used to analyze the result/response
   - Outcome of the analysis
   - Recommendation for the system/process

A detailed analysis by Montgomery[3] highlights the concept of experiment, with emphasis on non-statistical knowledge of the problem, use of simple design, and application of simple analysis, better to understand the different significance between practical, statistical, and semi-statistical experimental designs.

Chapter 2 discusses OFAT/OVAT followed by SDED. Chapter 3 gives detailed statistical experimental designs. Chapter 4 gives the analysis of selected experimental designs, followed by finding the optimal conditions of the biological process variables. Chapter 5 summarizes the concept of evolutionary operation programmes, whereas Chapter 6 analyses the robust experimental design of Taguchi. Chapter 7 overviews the concept of hybrid designs involving the genetic algorithm approaches.

The traditional experimental approach is to study the best guess approach and OFAT.[3]

2.2.1 OFAT/OVAT

If there are three factors in a biological system under investigation, one factor, namely variable A, will be varied on unspecified range, while the other two variables will be at some pre-determined fixed level. In this way, variable A will assume a level beyond which the desired response will not increase. For studying the effect of variable B only, variable A has a particular value obtained in the earlier experiment, whereas variable C is at pre-determined level *as* in the studies of the effect of variable A. In those experiments, variable B varies in an unexpected range like in the study of variable A to obtain a level beyond which the desired response will not increase further. In the third step, with the desired levels of variables A and B, variable C will take different levels similarly like the study of variables *A* and *B*. Then the desired level of *C* is determined from a rigorous set of experiments. In these three sets of experiments, the highest desired response value may not be same for the highest level of *A*, *B*, and *C* separately studied. If one uses the highest desired level of *A*, *B*, and *C* obtained by an earlier experiment to carry out a separate run, the response in this case again will be different.

Hence, the experimenters need to perform a large number of experiments at different levels along with the required replication. Finally, the experiments will not converge to a desired optimal condition. This value may be considered as suitable conditions/levels of variables *A*, *B*, and *C*. The following example will describe the process in detail.

Example 2.1

Three strains of *Saccharomyces cerevisiae*, namely, *A*, *B*, and *C*, were considered for electroporation studies. The amount of DNA required for transformation, pulse width (µs), field strength (kV/cm), growth phase, and salt concentration are major factors. OFAT approach is the only choice to the experimenter to find suitable conditions for electroporation to get higher transformants. However, the assumption is to study only two variables, that is, field strength and pulse width.

Solution:

In the first place, field strength varied between 2 and 5 (kV/cm) and pulse width of 5 ms kept at an arbitrary level. Figure 2.2 shows different suitable field strength for the three strains. In the next phase, separate runs were used for studying different levels of pulse width at a field strength of 2.75 kV/cm. Figure 2.3 summarizes the result.

Pulse width is different for different strains. Now, one can observe that for two factors, 12 separate runs are necessary. It does not give the best solution for a suitable value. For replication, the total number of experiments will be certainly high. This indicates more use of resources and time. One advantage is the freedom for the experimenter. Standard error and standard deviation can be calculated for these experiments.

2.2.2 Best Guess Approach

In a system/process, a few variables are assumed to influence the output (response). Based on initial guess and concept, the experimenter carries out an experiment. If the desired output has an appreciable improvement, the experimenter may stop with this guess without further taking a risk. If the experimenter is greedy for a further better result, the experimenter fine-tunes the previous change. Alternatively, if the first guess is disappointing, the experimenter looks for a change in the first choice and progresses to have a better result (Figure 2.4).

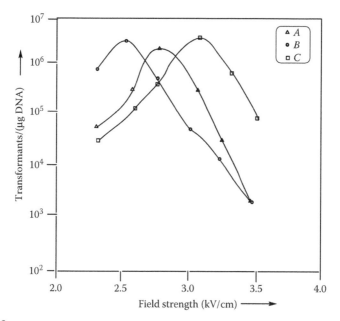

FIGURE 2.2
Field strength at constant pulse width to yield transformants.

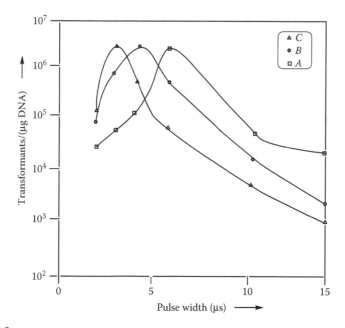

FIGURE 2.3
Effect of pulse width at constant field strength to yield transformants.

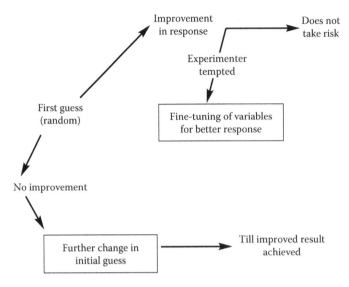

FIGURE 2.4
Best guess approach.

Drawbacks:

1. For no improvement in response, resource and time are consumed without guarantee.
2. For improvement in response, there is no guarantee to achieve best solution.

Who can venture this idea?

- Experienced scientists
- Practicing engineers
- Practicing technical manpower

Example 2.2

After studying the experiments, the experimenter guesses that salt composition in the recovery buffer used for recovery of cells after electroporation might increase the number of transformants. Three different salts, namely, X_1, X_2, and X_3, have been studied at different concentration.[4]

Solution:
Figure 2.5 shows that there is no improvement in the number of transformants at higher salt concentration. The guess is not correct. Therefore, one has to change the strategy of experiments. With respect to desired suitable values, there is no final answer from this experimental approach.

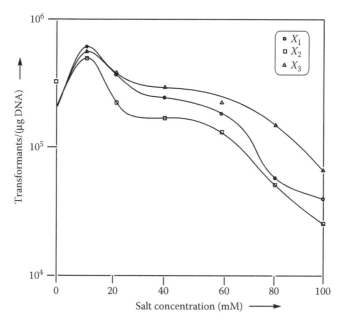

FIGURE 2.5
Effect of different salt concentration on the yield of transformants.

2.2.3 Self-Directing Experimental Design

SDED considers random combination of levels of selected variables for studying them simultaneously. This attempts to reduce the use of resource. Experiments of a few combinations of variables can be performed in a like-system under almost identical conditions in stipulated number of runs. After the evaluation of those runs, analysis will suggest for a next set of run with new combinations of variables. The initial number of runs is based on the geometric concept of presentation of the variables, which is called a *simplex*. The movement of the simplex directed by experiments in the right direction is the guideline for new run.[5]

2.2.3.1 Advantages

The following are the advantages of this technique:

No prior information is required between input variables and the response.

It is directed by stipulated experiments.

Mathematical postulation is not necessary.

Statistical significance tests are not necessary.

Simple experiments are implemented based on available analytical equipment.

Utilization of resource is very less when compared to OFAT/OVAT.

2.2.3.2 Overall Process

The following are the steps followed for the overall process:

Step 1: One needs to assume or pre-determine the variables in the process. This is the start of the experiment.

Step 2: Based on step 1, the experiments are performed to get the desired response.

Step 3: If the response is satisfactory, experiments are conducted in that direction to search for a domain containing the suitable values of the variables. This is guided by a defined geometric figure, called *simplex*.

Step 4: One needs to do one set of experiment at one time and wait for the result to do the next set of experiment, if it is necessary. The movement of a simplex in a right direction represents the entire process. The simplex is defined by an nth domain spatial arrangement by $(n + 1)$ points in a space. $(n + 1)$ is the number of vertices of the geometric figure having n number of variables. A similar kind of approach is reported by Hendrix.[6]

2.2.3.3 Theory of the Design

SDED is a non-statistical experimental design and does not require complex mathematical treatment. Levels of variables are determined by the experiments when simultaneous experimentation is not possible. Each vertex of a regular geometric figure, called the *simplex* corresponds to the conditions of an experimental run, that is, the run number. The simplex rotates in a preferred direction in the response plane. If an error occurs during the movement of the simplex, it self-corrects itself to the correct direction. Initially, the design stands with a selected set of experiments involving all variables. During the selected set of experiment, the worst result from the experiments is not considered but a new set of experiment is considered. The theory is the average of two times the best conditions minus the condition yielded lowest response. Again, after performing the experiment with the new conditions, the worst condition is identified and is replaced by a new set of conditions. This interactive procedure will continue to a direction when there is no improvement in response.

2.2.3.4 Limitations

The following are the limitations:

- Sequential experiments
- Replication increases time for decision-making process
- Important variables with their preferred levels are assumed without any proper justification

The theory is explained as follows:

- *Problem statement*: A, B, and C are three variables that influence the production of a metabolite.
- *Concept of a geometric figure*: Here, the geometric figure is a tetrahedron (Figure 2.6).

The movement is about the geometric centre of a tetrahedron in a response plane.

2.2.3.5 Initial Runs

The initial experimental conditions are tabulated in Table 2.1

Four separate experiments give four responses (*cf.* Table 2.2). On comparison, the run number giving worst response is identified, which needs to be discarded and a new experiment with a modified set of experimental conditions replaces the worst run.

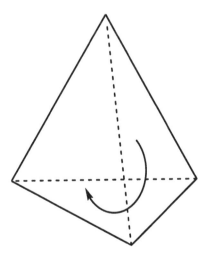

FIGURE 2.6
Tetrahedron.

TABLE 2.1

Initial Experimental Conditions

Variables ▶ Run Number ↓	A	B	C
1	L_A	H_B	H_C
2	L_A	H_B	L_C
3	H_A	L_B	L_C
4	H_A	L_B	H_C

Note: 'H' is the high level of a variable; 'L' is the low level of a variable.

TABLE 2.2

Response for Four Runs

Run Number	Final Response of the Experiment
1	$R1$
2	$R2$
3	$R3$
4	$R4$

TABLE 2.3

To Find Conditions for Next Run

Variables ▶ Run Number ↓	A	B	C
1	L_A	H_B	H_C
3	H_A	L_B	L_C
4	H_A	L_B	H_C
Sum of best points	$(L_A + 2H_A)$	$(H_B + 2L_B)$	$(2H_C + L_C)$
Average of best points (actual value to be indicated with measurable value)	$(L_A + 2H_A)/3$ $= A_{AV}$	$(H_B + 2L_B)/3$ $= B_{AV}$	$(2H_C + L_C)/3$ $= C_{AV}$
2 × average of best points	$2A_{AV}$	$2B_{AV}$	$2C_{AV}$
(2 × average of best conditions) − condition giving lowest response (in this case run number 2)	$(2A_{AV} - L_A)$	$(2B_{AV} - H_B)$	$(2C_{AV} - L_C)$
The combination of variable for run number 5	$(2A_{AV} - L_A)$	$(2B_{AV} - H)$	$(2C_{AV} - L_C)$

If $R2$ is the worst response, it is substituted by a new run #5. To get the combination of variables for run #5, the procedure is given as follows.

2.2.3.5.1 Calculation of Level of Variables for Run #5

Table 2.3 describes the procedure.

Experiment with the conditions of run #5 is carried out to get the response.

For iteration to get new run condition, runs #1, 3, 4, and 5 are considered and the response values are compared for them. There are two cases.

Case 1: If run #5 gives the worst response, there is no need to go in this direction. The suitable condition will lie in the highest response value among runs #1, 3, and 4. Then final decision will be taken for the process.

Case 2: If the run #5 gives better response than runs #1, 3, and 4, again the calculation for the new run is the same as described above. In this case, the worst response among runs #1, 3, 4, and 5 will be identified, and that run gave worst response will be rejected. Again the conditions for the new run will be evaluated as discussed in Table 2.3.

Example 2.3

In a batch bioreactor, an extracellular enzyme is produced by a fungal strain under controlled pH, airflow rate, and agitation rate. Run numbers are initially limited to four using different combination of variables as suggested in the plan. It appears that the process improves the desired product. Use the method of SDED and suggest the suitable operating condition for enzyme production.

Solution:

The method of Hendrix[6] and Panda et al.[5] have been employed to suggest the suitable combination of variables (Table 2.4).

1. Four separate runs are carried in the same reactor sequentially. Other conditions of production are maintained almost the same.
2. After certain period of fermentation, samples are collected to analyze for selected components, namely, enzyme activity, residual reactants, and cell mass. All of them or one of them will be the desired responses. In this case, extracellular enzyme activity is the desired response.
3. The fermentation continues for a specified time. In this case, it is 70 h.
4. Response along with the combination of variables is given in Table 2.5.
   Run #1 shows the most response among the four runs.
5. Then runs #2, 3, and 4 are considered to calculate probable levels of variables for the next new run. Table 2.6 summarizes the calculation.

Decision

Therefore, the levels of the variables in the new experiment are aeration rate = 1.8 (m³ air)/(m³ reaction medium) (min), agitation rate = 370 revolutions/min, and pH controlled at 3.30.

TABLE 2.4

Conditions for Initial Four Runs Carried Out Separately in a Batch Bioreactor

Variables ▶ Run Number ↓	Aeration Rate ([m³ air]/ [m³ Reaction Medium] [min])	Agitation Rate (rev/min)	pH (Controlled)
1	0.2 (*L*)	100 (*L*)	6 (*H*)
2	1.4 (*H*)	100 (*L*)	6 (*H*)
3	0.2 (*L*)	300 (*H*)	4 (*L*)
4	1.4 (*H*)	300 (*H*)	4 (*L*)

TABLE 2.5

Response at Different Fermentation Condition
Given in Table 2.4

Variables ▶ Run Number ↓	Response, Enzyme Activity
1	0.45
2	0.67
3	0.51
4	0.49

TABLE 2.6

Levels of Variables for the Next New Run

Variables ▶ Run Number ↓	Aeration Rate ([m³ air]/[m³ Reaction Medium] [min])	Agitation Rate (rev/min)	pH (Controlled)
2	1.4	100	6
3	0.2	300	4
4	1.4	300	4
Sum of best points (S)	3.0	700	14
Average of best points S/3 = AV	1.0	233.33 ≈233	4.66 ≈4.67
2 × AV	2.0	466	9.34
2 × AV – level of variable for worst run, that is, in this case, conditions of run #1	(2 – 0.2) = 1.8	(466 – 100) = 366 (≈370)[a]	(934 – 6.0) = 3.34 (≈ 3.30)[a]

[a] Approximation is due to the sensitivity of the measuring devices.

Exercises

2.1 Chitinase produced in a 5-L bioreactor in batch mode of operation at 30°C. Three parameters, namely, pH (controlled), aeration rate (m^3 air)/(m^3 reaction medium) (min), and agitation rate (revolution/min) were controlled at different levels. The experiments were planned as per SDED. Following information (Table 2.1.1) is available to make the decision for the process development.

Calculate the values of levels of variables as marked question mark above. Suggest the opinion for the process development.

2.2 For the production of gluconic acid, Table 2.2.1 gives initial four runs for the gluconic acid production. Suggest the production strategy.

Calculate the levels of variable for run #5, where the production of gluconic acid was 5.1 kg/m^3. Suggest the levels of variable for the proposed sixth run. Does it require proceeding in this direction further?

2.3 Table 2.3.1 suggests 10 experimental runs in a controlled batch bioreactor for the biological production of glutamic acid. The pH,

TABLE 2.1.1

Initial Results of SDED for Chitinase Production

Variables ▶ Run Number ↓	pH (Controlled)	Aeration Rate ([m³ air]/[m³ Reaction Medium] [min])	Agitation Speed (rev/min)	Response (U)
1	5.6 (H)	1 (L)	162 (L)	0.185
2	5.6 (H)	2 (H)	162 (L)	0.140
3	4.0 (L)	1 (L)	303 (H)	0.109
4	4.0 (L)	2 (H)	303 (H)	0.251
After the evaluation of the initial four runs,				
5	?	?	?	0.187

TABLE 2.2.1

Gluconic Acid Production Following SDED Procedure

Variables ▶ Run Number ↓	pH (Controlled)	Aeration Rate ([m³ air]/[m³ Reaction Medium] [min])	Agitation Speed (rev/min)	Response (kg/m³)
1	6	0.2	100	2.4
2	6	1.4	100	3.0
3	4	0.2	300	4.9
4	4	1.4	300	2.6

TABLE 2.3.1

Response with the Variables for 10 Runs in a Batch Bioreactor

Run Number	pH (Controlled)	Aeration Rate	Agitation Rate	Response, Glutamic Acid (kg/m³)
1	5	3	150	1.15
2	8	1	150	0.89
3	5	1	300	0.98
4	8	3	300	1.03
5	4	3.67	350	1.22
6	6.3	5.45	234	1.28
7	2.2	5.08	188	1.19
8	3.3	6.47	364	1.19
9	3	6.5	400	1.04
10	8	1	300	1.00

TABLE 2.4.1

Experimental Plan for Enzyme Production

Run Number	pH (Controlled)	Aeration Rate ([m³ air]/[m³ Reaction Medium] [min])	Agitation Speed (rev/min)	Response, Enzyme Activity (U)
1	6	0.2	100	0.451
2	6	1.4	100	0.661
3	4	0.2	300	0.508
4	4	1.4	300	0.489
5	?	?	?	0.649
6	?	?	?	0.910
7	?	?	?	0.870
8	?	?	?	0.895

aeration rate, and agitation rate are three important variables for the said fermentation. The experimenter casually fixed the range of variables. Suggest whether 10 runs are necessary for finding a suitable fermentation condition of those three variables. Consider the SDED procedure.

2.4 An enzyme production was studied in a batch bioreactor. You have only one complete set of bioreactor to find suitable conditions for controlled pH, aeration rate, and agitator speed.

    a. State the experimental plan you will adopt in this study.

b. Following data (Table 2.4.1) are available for the experimental plan.

    i. Find the unknown values in Table 2.4.1.

    ii. What procedure will you follow?

    iii. Which are the suitable conditions?

References

1. Myers RH and Montgomery DC, *Response Surface Methodology*, 2nd edition, John Wiley & Sons, 2002.
2. Box GEP, Choice of response surface design and aliphatic optimality, *Utilitas Mathematica*, **21B**, 11–55, 1982.
3. Montgomery DC (Ed.), *Design and Analysis of Experiments*, 7th edition, Wiley, India, 2009.
4. Lakshmi Prasanna G, Studies on electrotransformation of *S. cerevisiae* and electrofusion of *S. cerevisiae* and *Trichoderma reesei*, PhD Thesis, Indian Institute of Technology, Madras, India, 1999.
5. Panda T, Babu PSR, Kumari JA, Rao DS, Théodore K, Rao KJ, Sivakesava S et al., Bioprocess optimization – A challenge, *Journal of Microbiology and Biotechnology*, 7(6), 367–372, 1997.
6. Hendrix C, Through the response surface with test tube and pipe wrench, *Chemtech*, **10**, 488–497, 1980.

Further Reading

Dasu VV, Studies on production of griseofulvin by *Penicillium griseofulvum*, PhD Thesis, Indian Institute of Technology, Madras, India,1999.

Lakshmi Prasanna G, Studies on electrotransformation of *Saccharomyces cerevisiae* and *Trichoderma reesei*, PhD Thesis, Indian Institute of Technology, Madras, India, 1999.

Naidu GSN, Studies on behavior and production of extracellular pectinases from *Aspergillus niger*, PhD Thesis, Indian Institute of Technology, Madras, India, 1999.

Théodore K, Studies on optimization of β-1,3-glucanase production by *Trichoderma harzianum* NCIM 1185, PhD Thesis, Indian Institute of Technology, Madras, India, 1995.

3

Response Surface Experimental Designs

OBJECTIVE: A detailed experimental plan before conducting the experiment is the basis of experimental design. The design is selected after setting the objectives and selecting the process variables. The concept of different models and experimental designs is explained with reference to biological processes.

3.1 Introduction

Response surface approach, a sequential search, is a combination of mathematical and statistical tools to find empirical functional relations between true mean response and a number of input variables (Figure 3.1). For biological processes, the process engineer wants to know the values of pH ($\equiv x_1$) and temperature ($\equiv x_2$) to obtain maximum yield (Y) of a particular metabolite. x_1 and x_2 are the controlled variables.

Therefore, $Y = f(x_1, x_2) + \text{Error}$.

The surface of the response is expressed by $\mu = f(x_1, x_2)$.[1]

The fitted surface is used for this analysis. If one assumes that the fitted surface approximately represents true response, then the analysis of fitted surface will be equivalent to the actual system. The evaluation of model parameters will be appropriate, if proper experimental design in used to gather the information. Those experimental designs are called *response surface design*.[1]

3.2 Principal Objective of Response Surface Method

- To determine optimum operating values of controlled variables
- To determine a space for factor/controlled variable where operating conditions are satisfactorily achieved.

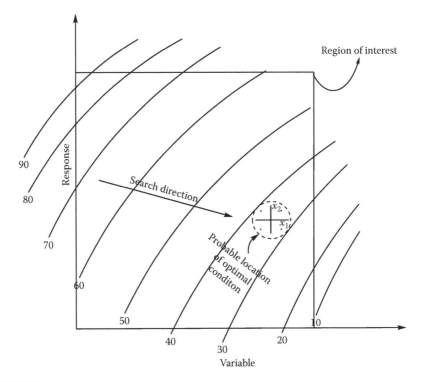

FIGURE 3.1
Schematic representation of response surface experimental searches.

3.3 Drawback

In this technique, no proper relationship exists between the response variable and the controlled variables.

3.4 Types of Response Surfaces

Johnson and Leone[1] suggested that the fitted surface may be one of the following types.

- A possible maximum – It is defined in Figure 3.2. The condition is $(\partial\mu/\partial x) = 0$.
- A possible minimum – The preferred condition is $(\partial\mu/\partial x) = 0$. Figure 3.3 gives the concept.

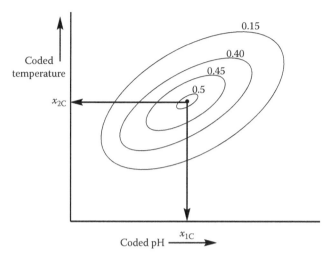

FIGURE 3.2
Representation of a possible maximum.

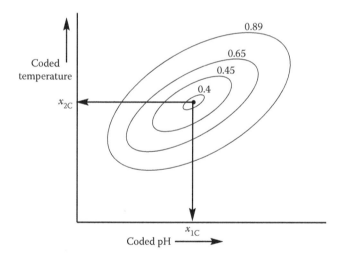

FIGURE 3.3
Representation of a possible minimum.

- Saddle point – It is difficult to attain optimal value by the solution of $(\partial\mu/\partial x) = 0$ (Figure 3.4).
- Rising ridge – This concept is given in Figure 3.5.
- Stationary ridge – Figure 3.6 is the example of a stationary ridge.

In Figure 3.5, the optimal lies on the broken line, whereas in Figure 3.6, there are many conditions satisfying optimal requirement.

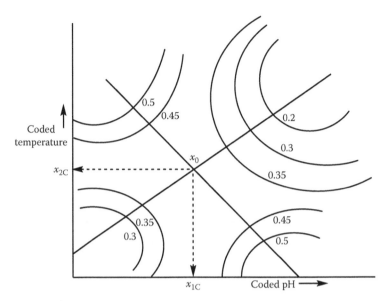

FIGURE 3.4
Representation of saddle point.

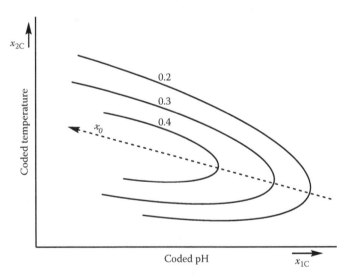

FIGURE 3.5
Representation of rising ridge.

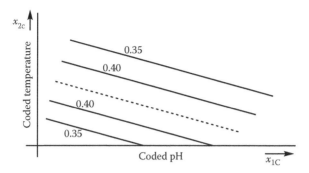

FIGURE 3.6
Representation of stationary ridge.

3.5 Classification of Response Surface Designs

Figure 3.7 represents the possible classes of response surface designs.

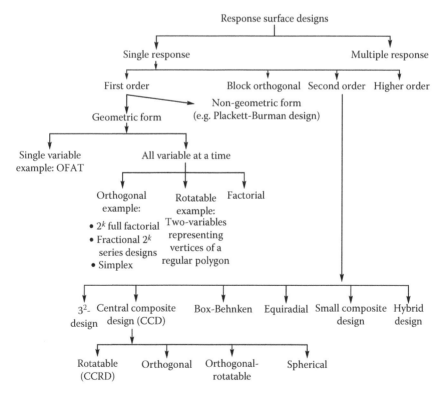

FIGURE 3.7
Classification of response surface designs.

3.6 First-Order Designs

The first-order response surface model is Equation 3.1.

$$\tilde{Y} = \beta_0 + \beta_1 x_1 + \cdots + \beta_k x_k + \varepsilon \tag{3.1}$$

where:
$\tilde{Y}$ is the predicted response
β_0 is the intercept estimate
β_k is the estimate of unknown coefficient for k_{th} factor

Therefore, the experimental designs from which β_k terms (linear effect terms) can be estimated are known as *first-order design*.

They are of two types based on the design points representing geometric figure or not. In 2^k-full and fractional factorial designs and simplex design, the design points represent the geometric figure. On the other hand, the design points in Plackett-Burman design[2] do not represent the geometric figure; they are non-geometric design.

Again, the first-order geometric designs are subdivided into one factor at a time (OFAT) and all variables at a time. All variables at a time is of orthogonal and rotatable designs.

3.6.1 OFAT/One Variable at a Time

Koshal design[3] is the one variable at a time (OVAT). This design is without and with interaction of variables. This is an example of saturated design. However, saturated design does not allow estimation of error. Care is necessary for using this design. For this reason, at least one centre point appears in the design matrix.[4] The following are the design matrices for Koshal designs of the first-order model.

Matrix for OFAT with no interaction

$$D = \begin{vmatrix} 0 & 0 & 0 & 0 \\ 0 & 0 & 0 & 1 \\ 0 & 0 & 1 & 0 \\ 0 & 1 & 0 & 0 \\ 1 & 0 & 0 & 0 \end{vmatrix}$$

Design matrix for OFAT with a two-factor interaction

$$
D = \begin{vmatrix}
0 & 0 & 0 & 0 \\
0 & 0 & 0 & 1 \\
0 & 0 & 1 & 0 \\
0 & 1 & 0 & 0 \\
1 & 0 & 0 & 0 \\
1 & 0 & 0 & 1 \\
1 & 0 & 1 & 0 \\
1 & 1 & 0 & 0 \\
0 & 0 & 1 & 1 \\
0 & 1 & 1 & 0 \\
0 & 1 & 0 & 1 \\
0 & 1 & 0 & 1
\end{vmatrix}
$$

This design may be extended to three-factor interaction. Following is the matrix:

$$
D = \begin{vmatrix}
0 & 0 & 0 & 0 \\
0 & 0 & 0 & 1 \\
0 & 0 & 1 & 0 \\
0 & 1 & 0 & 0 \\
1 & 0 & 0 & 0 \\
1 & 0 & 0 & 1 \\
1 & 0 & 1 & 0 \\
1 & 1 & 0 & 0 \\
0 & 0 & 1 & 1 \\
0 & 1 & 1 & 0 \\
0 & 1 & 0 & 1 \\
1 & 1 & 1 & 0 \\
1 & 1 & 0 & 1 \\
1 & 0 & 1 & 1 \\
0 & 1 & 1 & 1
\end{vmatrix}
$$

Koshal design appears to be suitable for OFAT mentioned in Chapter 2. Biological systems can be effectively analyzed with this design, as it suggests suitable interactions of variables in the first-order model. Variables should be limited to defined levels. This could be one of the differences between the experimental approaches explained in Chapter 2 for OFAT and the procedure of Koshal design limited to the first-order model.

3.6.2 All Variables at a Time

The discussion refers to orthogonal first order and rotatable first-order designs.

1. *Orthogonal first-order designs*

   The condition defined by Montgomery[2] is off-diagonal elements of $(X'X)$ matrix is all zero.

2. *Rotatable first-order designs*

   The designs in which $V(x_1, x_2, ..., x_k)$ is a function of $\sum_{i=1}^{k} x_i^2$. Johnson and Leone[1] suggested that the variance of ή $(x_1, x_2, ..., x_k)$ depends on the distance of the point $(x_1, x_2, ..., x_k)$ from the point $(0, 0, ..., 0)$. The variance depends on the distance if the design rotates though an arbitrary angle about the point $(0, 0, ..., 0)$. Therefore, these designs are rotatable. Figure 3.8 presents the first-order geometric rotatable design.

3. *Factorial experimental design*

   To vary several factors at the same time is the correct approach for a biological experiment compared to the OFAT/OVAT method of design. This can also be divided into following categories (Figure 3.9).

Figures 3.10 and 3.11 describe the classes of factorial experimental designs.

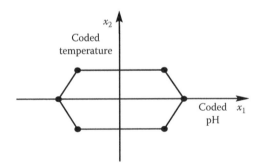

FIGURE 3.8
First-order geometric rotatable designs.

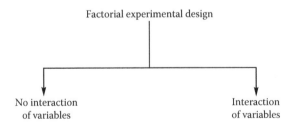

FIGURE 3.9
Classification of factorial experimental designs.

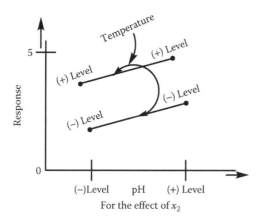

FIGURE 3.10
No interaction of variables.

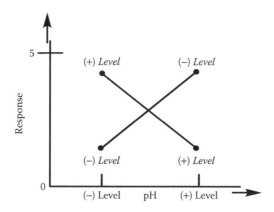

FIGURE 3.11
Interaction of variables.

First-order designs follows first-order model to predict the response. The first-order model contains linear terms. The curvature in the contour is due to the interaction terms. For this reason, 2^2-designs are the first order-designs, whereas 3^2-designs are the second-order design.

3.6.2.1 2^k Factorial Experimental Designs

Geometrically, the conditions for each run are the vertex of the geometric figure, where k is the number of variables. For example, $k = 2$, and $2^k = 4$. There are four combinations of variables (Figure 3.12). Out of at least six probable combinations, let us consider one of the combinations for a detailed analysis of effect of factors (Figure 3.13). All 2^k designs are orthogonal design.[4]

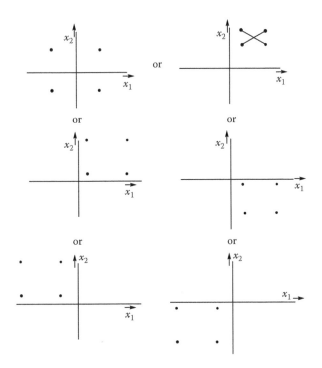

FIGURE 3.12
Various options of 2²-design.

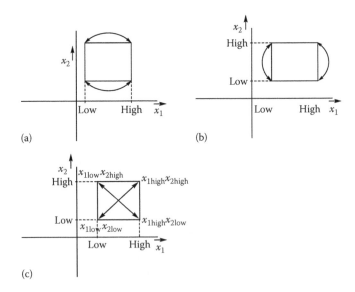

(a)

(b)

(c)

FIGURE 3.13
Detailed analysis of effects of factors in 2²-designs: (a) for the effect of x_1, (b) for the effect of x_2, and (c) for the interaction effect of x_1 and x_2.

3.6.2.1.1 2^2-Design

Steps: Table 3.1 is an example of completely randomized experiment for a detailed analysis.

Figure 3.14 represents 2^2-factorial designs points.

$x_{1i}, x_{2i}, x_{3i},$ and x_{4i} are the total responses for four individual runs. Each run has three replicates. For each run, three responses are obtained for the replicates, that is, $x_{11}, x_{12},$ and x_{13}. Hence, $x_{11}, x_{12},$ and x_{13} contribute for one total response, x_{1i}. Similarly, $x_{2i}, x_{3i},$ and x_{4i} are obtained. Here, three replicates for each run give one total response. Generalization of number of replicates for each run gives the total responses for n replicates.

Calculation of main effect of x_1

$$\text{Main effect } (x_1) = \frac{\begin{pmatrix} \text{Effect of } x_1 \text{ at high level of } x_2 + \\ \text{Effect of } x_1 \text{ at low level of } x_2 \end{pmatrix}}{2}$$

$$\text{Effect of } x_1 \text{ at high level of } x_2 = \frac{(x_{4i} - x_{3i})}{\text{Number of replications}}$$

TABLE 3.1

Analysis of 2^2-Factorial Designs

Run Order	Variables with the Condition		Response at Three Replications			Total Response
	x_1	x_2	1	2	3	
1	x_{1L}	x_{2L}	x_{11}	x_{12}	x_{13}	x_{1i}
2	x_{1H}	x_{2L}	x_{21}	x_{22}	x_{23}	x_{2i}
3	x_{1L}	x_{2H}	x_{31}	x_{32}	x_{33}	x_{3i}
4	x_{1H}	x_{2H}	x_{41}	x_{42}	x_{43}	x_{4i}

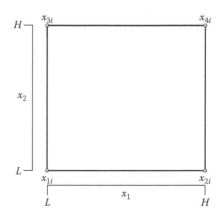

FIGURE 3.14

Presentation of data in Table 3.1 pertaining to 2^2-factorial design points.

$$\text{Effect of } x_1 \text{ at low level of } x_2 = \frac{(x_{2i} - x_{1i})}{\text{Number of replications}}$$

$$\text{Therefore, the main effect of } x_1 = \frac{(x_{4i} - x_{3i}) + (x_{2i} - x_{1i})}{2 \times \text{Number of replications}}$$

$$\text{Similarly, the main effect of } x_2 = \frac{(x_{4i} - x_{2i}) + (x_{3i} - x_{1i})}{2 \times \text{Number of replications}}$$

Interaction effect of x_1 and $x_2 =$

$$\frac{[(\text{Effect of } x_1 \text{ at high level of } x_2) - (\text{Effect of } x_1 \text{ at low level of } x_2)]}{2 \times \text{Number of replications}}$$

or

Interaction effect of x_1 and $x_2 =$

$$\frac{[(\text{Effect of } x_2 \text{ at high level of } x_1) - (\text{Effect of } x_2 \text{ at low level of } x_1)]}{2 \times \text{Number of replications}}$$

$$\frac{(x_{4i} - x_{2i}) - (x_{3i} - x_{1i})}{2 \times \text{Number of replications}}$$

which is the above expression as well. There are alternative methods to calculate the effects. However, the result is the same.

Alternative method for the calculation of main effects and interaction effect

Main effect $(x_1) =$

(Average response of two-treatment combination of x_1 at higher level)

$-$ (Average response of two-treatment combination of x_1 at lower level)

$$= \frac{(x_{4i} + x_{2i})}{2 \times \text{Number of replications}} - \frac{(x_{3i} + x_{1i})}{2 \times \text{Number of replications}}$$

$$= \frac{(x_{4i} + x_{2i} - x_{3i} + x_{1i})}{2 \times \text{Number of replications}}$$

This is equally true for the main effect of (x_2) and the interaction effect of $(x_1$ and $x_2)$.

The statistic required for this analysis is described below.

Contrast

The contrast of $x_1 = \text{Cont } x_1 = (x_{4i} + x_{2i} - x_{3i} - x_{1i})$

This is called *total effect* of x_1. Similarly, the Cont x_2 and Cont x_1x_2 are calculated. Sum of squares is calculated based on the formula of Montgomery.[1]

$$\text{Sum of squares} = \frac{\text{Contrast square}}{\left(\begin{array}{c}\text{Number of observations in}\\\text{each total in the contrast}\end{array}\right) \times \left(\begin{array}{c}\text{Sum of square of the}\\\text{contrast coefficient}\end{array}\right)}$$

In this example,

$$SSx_1 = \frac{\left(\text{Cont}\,x_1\right)^2}{4 \times \text{Number of replications}}$$

$$SSx_2 = \frac{\left(\text{Cont}\,x_2\right)^2}{4 \times \text{Number of replications}}$$

$$SSx_1x_2 = \frac{\left(\text{Cont}\,x_1x_2\right)^2}{4 \times \text{Number of replications}}$$

Total sum of squares (SST) and error sum of squares (SSE) are also calculated, where SSE and SST are correlated based on the correlation suggestions of Montgomery.[2]

$$SSE = SST - SSx_1 - SSx_2 - SSx_1x_2$$

Table 3.2 summarizes the result.

TABLE 3.2

ANOVA

Source of Variation	Sum Squares	Degrees of Freedom	Mean Square	F_{Value}
x_1	SS_{x_1}	1	$MS_{x_1} = \dfrac{SS_{x_1}}{d.f.\ associated\ with\ SS_{x_1}}$	$\dfrac{MS_{x_1}}{MSE}$
x_2	SS_{x_2}	1	$MS_{x_2} = \dfrac{SS_{x_2}}{d.f.\ associated\ with\ SS_{x_2}}$	$\dfrac{MS_{x_2}}{MSE}$
x_1x_2	$SS_{x_1x_2}$	1	$MS_{x_1x_2} = \dfrac{SS_{x_1x_2}}{d.f.\ associated\ with\ SS_{x_1x_2}}$	$\dfrac{MS_{x_1x_2}}{MSE}$
Error	SSE	(4 × Number of replications − 1) − 3	$\dfrac{SSE}{(4 \times \text{Number of replications} - 1) - 3}$	
Total	SST	(4 × Number of replications − 1)		

Note: For the example given in Table 3.1, treatment combinations are (x_{1L}, x_{2L}); (x_{1H}, x_{2L}); (x_{1H}, x_{2H}); and (x_{1H}, x_{2H}). Therefore, there are four treatment combinations. Each treatment combination has three replicas. Degrees of freedom associated with SSE calculation is given by

{(number of treatment combinations, that is, in this case, it is 4) × (number of replicas, that is 3)} − 1 − sum of degrees of freedom associated with SS_{x_1}, SS_{x_2}, and $SS_{x_1x_2}$.

Similarly, degrees of freedom for SST is calculated.

Regression model

The regression model of a 2^2-design in the example is given by Equation 3.2.

$$\tilde{Y} = \beta_0 + \beta_1 x_1 + \beta_2 x_2 + \varepsilon \tag{3.2}$$

x_1 and x_2 are coded variables defined by Montgomery.[2]

$$x_i = \frac{X_i - \left(X_{ilow} + X_{ihigh}\right)/2}{\left(X_{ihigh} - X_{ilow}\right)/2} \tag{3.3}$$

and $i = 1$ and 2.

The response surface can be generated from the fitted model Equation 3.2 (*cf.* Figure 3.15).

Figure 3.15 is the result of assembling the first-order model, which contains only the main effects of the variables.

Variation in factorial design

Following specific examples may be considered here.

- Centre points in 2^k-design
- Block in 2^k-design
- Confounded in 2^k-design
- Fractional factorial design
- Specific 3^k-design

Centre points in 2^k-designs

In a 2^2-design, the addition of centre points quantifies the curvature in the response surface (Figure 3.16).

The following is the model equation suggested by Montgomery.[2]

$$\tilde{Y} = \beta_0 + \sum_{j=1}^{k} \beta_j x_j + \sum_{i<j} \sum \beta_{ij} x_i x_j + \sum_{j=1}^{k} \beta_{jj} x_j^2 + \varepsilon \tag{3.4}$$

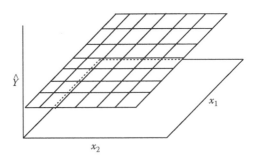

FIGURE 3.15
Response surface.

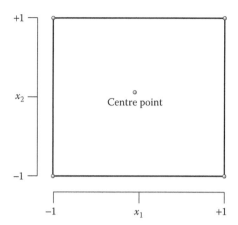

FIGURE 3.16
Centre point in a 2^2-design.

The extension of this design is the central-composite design (CCD), which will be discussed in the second-order design.

Example 3.1

The effect of initial pH of the enzyme production medium and temperature have been studied in a batch bioreactor on the production of extracellular enzyme. The experimental plan along with the response for separate runs is shown in Table 3.3. Write the coded values of the variables. Give the analysis of this 2^2-design.

Solution:

Code of pH: $pH_{low} = 5$; $pH_{high} = 7$
Code for temperature: $T_{low} = 25$, $T_{high} = 35$

Analysis:
There is no centre point. Decoded and coded variables are in Tables 3.3 and 3.4, respectively. The procedure for analysis is described in Section 3.6.2.1.1.

TABLE 3.3

Experimental Plan with Response

Run No.	Initial pH X_1	Temperature (°C) X_2	Chitinase (U) Experimental
1	5	25	0.1235
2	7	25	0.0301
3	5	35	0
4	7	35	0

TABLE 3.4

Coding Form of Table 3.3

Run No.	Initial pH X_1	Temperature (°C) X_2	Equivalent Point in the Figure 3.14	Chitinase (U) Experimental
1	Low	Low	(1)	0.1235
2	High	Low	a	0.0301
3	Low	High	b	0
4	High	High	ab	0

The main effect of temperature: ½ $(0 + 0 - 0.0301 - 0.1235) = -0.0768$

Main effect of pH: ½ $(0 + 0.0301 - 0 - 0.1235) = -0.0467$
Interaction effect of pH and temperature: ½ $(0 + 0.1235 - 0.0301 - 0)$
$= 0.0467$

The effect of pH is higher than the effect of temperature, which suggests that the initial pH of the medium from low value to high value will influence more. A change in temperature of fermentation from low value to high value will have a drastic effect on the enzyme activity. Interaction effect is more, which suggests strong interaction between pH and temperature.
Total effect or contrast: for pH is $= (0 + 0.0301 - 0 - 0.1235) = -0.0934$.
Other contrasts are calculated in a similar way.

Calculation of SS:

$$SS_{pH} = \frac{(-0.0934)^2}{4}$$

$$SS_{temperature} = \frac{(-0.1536)^2}{4}$$

$$SS_{pH-temperature} = \frac{0.0934^2}{4}$$

Now, SST is possible to calculate. Then, $SSE = SST - SS_{pH} - SS_{temperature} - SS_{pH-temperature}$.
Table 3.2 gives the procedure for the complete analysis of ANOVA for this problem.

This problem can be modified for other varieties of 2^k design.

Block in 2^k-designs

It is not always possible to run all the experiments with one-resource components even with the factorial points. If there are variations in the resource input quality, there will be differences in the response. For example, fermentation runs require malt extract. The same quality of malt extract may not be available for all the runs. This will affect the response. This effect is called *nuisance factor*. To understand the impact of the *nuisance factor* on the response, the experiments are carried out in blocks.

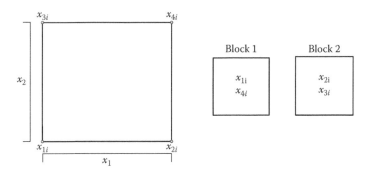

FIGURE 3.17
A 2^2-design in two blocks.

When a response surface design experiments are divided into blocks (*cf.* Figure 3.17) without affecting the estimation of parameters in the response surface model, then it is called *orthogonally blocked design*.

Application: The above procedure reduces the effect of noise on the response.

When to do blocking experiments?

This example gives an idea to understand blocking effect. However, if only one variable is the major source of variation, it is not wise to do blocking experiments. If, except the variable being *nuisance*, other factors not included in the experiment affect the performance, it is better to do blocking experiments. In blocking experiments, the number of runs will be more, which will increase precision and reduce confidence intervals.

Confounding 2^k-designs

Let us assume that cell concentration is measured by absorbance (method 1) and cell dry weight (method 2) in the fermentation broth. Cells are produced in two different medium compositions, namely, A and B. After data analysis, it is not possible to distinguish the effect of methods of cell measurement and the effect of cell production.

Therefore, the cell measurement technique and the production medium for cells have been confounded.

Rule for construction of blocks

Johnson and Leone[1] suggested the rule for construction of blocks:

- The block containing treatment (1) is the principal block.
- The principal block contains an even or a zero number of letters common with *confounding interaction*.
- For other blocks, experimenter has to select a treatment combination, which is not in principal block.

Example 3.2

For 2^3-factorial design, the factor *abc* is confounded. The experiments were carried out in two blocks.

Solution:
In 2^3-factorial design, eight experiments are necessary. In the laboratory, it is not possible to carry out eight experiments simultaneously. A maximum of four experiments can be carried out as per plan in each of the two blocks (see Figure 3.18) In block 1, they are (1), ab, ac, and bc, and for block 2, a, b, c, and abc.

In this example, the principal block has treatment (1), which has zero letters common with confounded ABC. *ab*, *bc*, and *ca* have two letters common with confounded ABC. There are no other possibilities.

For other blocks, treatments are provided in Table 3.5.

Example 3.3

In Example 3.2, assume C is confounded. Then write the treatments in the blocks.

Solution:
Figure 3.19 is the answer.

Fractional factorial designs

The example of 2^2-full-factorial-design requires four runs. It is appreciable to carry out all four experiments with replicates. Now, for 2^3-full-factorial-design, eight experiments are necessary. With replication, the numbers of experiment are more. Let us take 2^6-full-factorial-design. It requires 64 runs. With replicates,

Block1	Block2
(1)	a
ab	b
ac	c
bc	abc

FIGURE 3.18
2^3-full factorial design points.

TABLE 3.5

Treatments in a Block

$a \times (1) = a$
$a \times ab = a2b = b$
$a \times bc = abc = abc$
$a \times ca = a2c = c$

Block1	Block2
(1)	c
a	ac
b	bc
ab	abc

FIGURE 3.19
Factor 'C' confounded in the experiment.

it is difficult to carry out all the experiments. This may be due to limitation of resources, cost, and facilities to do experiments with equal care.

Therefore, it is necessary to select the portion of the experiments of the full factorial designs. This gives an idea of fractional factorial designs. Screening experiments consider fractional factorial points, where the experiments need to select only major factors influencing greatly on the response of the process.

Classification of fractional factorial designs of 2^k

They are as follows:

- *One-half fraction*, namely, $1/2 \times 2^k$ (*cf.* Figure 3.20)

 For full factorial of 2^3-designs, experimental points are given in Table 3.6.

 It is better to calculate effects of each component in Table 3.6. Montgomery[2] suggests ABC as *generator component*. In Table 3.6, there are only four alias pairs whose effects are equal. Therefore, there is no reason to carry out all eight experiments. Four experiments will be enough in this case.

 Hence, the 2^3-design is reduced to $1/2 \times 2^3$-design. This is called *one-half fraction design*. Effects are calculated as described in Section 3.6.2.1.1

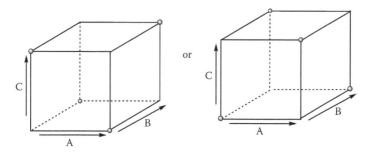

o→ Represents preferred experimental points

FIGURE 3.20
Representation of $\frac{1}{2} \times 2^3$ design.

TABLE 3.6

Components in 2^3 Designs

A	BC
B	AC
C	AB
1 = identify	ABC = All interaction

In this example:

The effect of A = the effect of BC

The effect of B = the effect of AC

The effect of C = the effect of AB

The effect of ABC = the effect of *1*

On this basis, one can find the $1/2 \times 2^4$, $1/2 \times 2^5$ designs, $1/2 \times 2^k$-designs.

- *Resolution III, IV, and V designs*

Montgomery[2] classifies the half-fractional design as Resolution III, IV, and V designs.

The generalized design is 2^{k-r}, where $r < k$. All the experiments are in one block. However, if experiments are in different blocks, the number of blocks ($=b$) are included in the total number of experiments. Number of experiments = $2^{(k-r-b)}$.

The two-level fractional factorial designs give rise to the concept of Plackett-Burman design.[5]

3.7 Non-Geometric Design

3.7.1 Plackett-Burman Design

This is a two-level fractional factorial design, but the design points cannot be presented as cubes.[5] Therefore, this design is called a *non-geometric design*.[5] However, the characteristics of this design are as follows:

Total number of runs (N) is equal to number of variables (k) plus one, that is $N = (k+1)$, provided following conditions are satisfied:

- N is a multiple of 4.
- They are orthogonal designs.
- This is two-level fractional-factorial designs. Simultaneously, it is called a *non-regular designs*.
- Some factor is partially aliased.
- This is a screening design, that is, if there are number of variables, some of the variables will have least effect on the responses. Those variables can be separated from the important variables. Important variables will be optimized in the second-order design.
- Then it follows the preparation of proper design matrix. This is for k.

3.7.1.1 Construction of the Plackett-Burman Design

To write a Hadamard matrix of the following constituents, the procedure is for k variables.

1. If the number of variables is not sufficient to form a Hadamard matrix as per in Plackett-Burman design, dummy variables are added to meet the criterion. For example, in a process, the number of variables for screening experiment is 10. Then $N = 10 + 1 = 11$, which is not a multiple of four. To obtain a nearest number multiple of four is 12. Hence, one dummy variable is necessary in this case.

2. Variables are coded +1 or –1. Low level of a variable will be from supported literature or from preliminary experiment. Higher level of the variable may be 25% more than the low level. This value depends on the region of experiment.

3. The number of positive ones is $(k+1)/2$ and the negative ones is $(k–1)/2$. They are not fractional number.

4. Terms stated in (3) is the composition of first row.

5. The next $(k–1)$ rows come from the first row by cyclical shifting of one place $(k–1)$ times, that is, the second row is obtained from the first row by shifting it one place. The third row is obtained from the second row by shifting it one place until the last but one row.

6. Finally, elements in the last row is of negative one. Total number of runs $= N = (k+1)$.

7. Note that there is no row where all elements are positive.

8. This gives an orthogonal design. The terms in the fitted model, to correlate linear components to the response, are uncorrelated with one another. Thus, parameter estimate are uncorrelated.

9. If N is a power of 2 and $N > (k+1)$, then this orthogonal design is altered to fractional factorial designs. N is not necessarily a power of 2, but it should be divisible by four.

3.7.1.2 Method to Perform Experiments Using the Plackett-Burman Design

Let us take one example. The medium used for fermentation contains glucose, Na_2SO_4, $MgCl_2$, peptone, citric acid, and $FeCl_3$; they are designated as variables. Among them, not all of the variables are influencing in the same manner. Some of them highly influence the fermentation process. It is, therefore, necessary to screen the variables.

The objective of screening is to reduce the number of variables, which will be used for optimization studies. For optimization experiments, if one encounters with large number of variables, it is difficult to achieve an optimal solution. The process is as follows:

Step 1—Coding the variables: Two levels of each variable are designated as +1 and –1 for higher value and lower value, respectively.

Step 2—Abbreviation of variables: The following are the five chemical compounds for medium formulation:

Glucose = A

Na_2SO_4 = B

$MgCl_2$ = C

Peptone = D

Citric acid = E

$FeCl_3$ = F

Step 3—Levels of variables: Levels are either from available literature or from initial trial runs or from experience in biological processes (Table 3.7).

Step 4—Formulation of Hadamard matrix: Hadamard matrix satisfies the criteria of Plackett-Burman design. To write the Hadamard matrix, text given in Section 3.7.1.1 is followed systematically.

For (a) total number of runs: $N = (k+1)$, which will be divisible by four.

For the present example, $N = (6+1) = 7$. This is not divisible by four. Hence, one dummy variable is included to form the matrix.

In this case, the dummy variable is G. This has also two levels +1 (or simply +) and −1 (or −)levels. G is an imaginary variable, which will appear in the matrix formulation, but it will not have any participation in the medium formulation. Hence, the total number of variables, including the dummy variable, is seven. $N = (7+1) = 8$, which satisfies the conditions for the Plackett-Burman design.

For (b) elements of first row: The number of positive ones = $(7+1)/2 = 4$.

The number of negative ones = $(7–1)/2 = 3$. They are not fractional quantities.

For each variable, in each column, the total number of positive ones and negative ones are same. Therefore, the experiments will have separate condition of every run. In this case, every run gives the composition of medium constituents. Eight different medium compositions are obtained from the Hadamard matrix (Table 3.8).

TABLE 3.7

Values of Variables

Variable	Low Level (kg/m³)	High Level (kg/m³)
A	5 ($\equiv$ −1)	25 ($\equiv$ +1)
B	0.1 ($\equiv$ −1)	1.0 ($\equiv$ +1)
C	0.05 ($\equiv$ −1)	1.5 ($\equiv$ +1)
D	0.5 ($\equiv$ −1)	2.0 ($\equiv$ +1)
E	1.4 ($\equiv$ −1)	3.0 ($\equiv$ +1)
F	0.005 ($\equiv$ −1)	0.1 ($\equiv$ +1)

TABLE 3.8

Hadamard Matrix for Cyclically Anticlockwise Rotation of Rows

Run Number	Variables in Coded Form						
	A	B	C	D	E	F	G
1=(1)	+	+	+	+	−	−	−
2=(7)	+	+	+	−	−	−	+
3=(6)	+	+	−	−	−	+	+
4=(5)	+	−	−	−	+	+	+
5=(4)	−	−	−	+	+	+	+
6=(3)	−	−	+	+	+	+	−
7=(2)	−	+	+	+	+	−	−
8=(8)	−	−	−	−	−	−	−

Figure in parentheses ( ) is the run number for cyclically clockwise arrangement of the matrix.

Step 5—Experimentation: One needs to carry out the experiments in flask culture because of eight different runs; with replicates, it is possible to run simultaneously all the experiments. Eight different medium compositions are prepared in eight separate flasks, at least. For replicates, more number of flasks is required to be prepared. This will be followed by sterilization and fermentation at suitable conditions. Samples have been withdrawn from all flasks regularly to estimate the desired response parameter, namely, the desired product concentration or cell concentration.

Step 6—Data collection: Data at maximum level of response are recorded, which are included in the Hadamard matrix.

Step 7—Computation: After obtaining the results (response), the following data are computed:

i. Effect of each variable =

$$\frac{\left(\text{Response at} + 1\,\text{level}\right)}{\left(\text{Number of runs}\right)/2} - \frac{\left(\text{Response at} - 1\,\text{level}\right)}{\left(\text{Number of runs}\right)/2}$$

ii. Mean effect $= \dfrac{\sum \text{Response}}{\text{Number of runs}}$

iii. t-value $= \dfrac{\text{Effect of variable}}{\text{Standard error}}$

iv. Probability $> |t|$

v. Confidence level $= (1 - \text{Probability due to chance}) \times 100\%$

Step 8—Response equation and regression analysis: Equation 3.1 contains only linear terms.

$$\tilde{Y} = \beta_0 + \beta_1 x_1 + \cdots + \beta_k x_K + \varepsilon \tag{3.1}$$

Coefficients in first-order model equations are β_0's and β_i's (Appendix A.2).

Step 9—Decision: It is assumed that the variables having confidence levels 80% and above are the important variables. Those variables will be optimized in the second stage of the experimental design. The confidence level cut off to 80% is subjective and will vary depending on the number of variables screened.

Example 3.4

Medium components were screened for extracellular enzyme production by a fungus in shake culture. Fermentation conditions were volume of the enzyme production medium – 100 cm³ in 500 cm³ Erlenmeyer flask. Constituents of the unoptimized medium (kg/m³) were major carbon source 10 and other medium constituents were added as given in Table 3.9. Other conditions were age of the organism – 120 h, inoculums age –36 h, inoculums level – 5% (v/v) (0.24 g/l dry equivalent of cells approximately), shaker speed – 180 rpm, fermentation time – 216 h, temperature of fermentation – 30°C, initial pH of the enzyme production medium – 5, and pH uncontrolled during fermentation. Results shown are the average of maximum endoglucanase production in duplicate experiments. Write the Hadamard matrix for the variables mentioned in Table 3.9.

Solution:

1. Coding of variables is as per Table 3.10.
2. In this case, three dummy variables appear to satisfy the condition of the Plackett-Burman experimental design.

TABLE 3.9

Other Medium Constituents for Enzyme Production

S. No.	Variables
A	K-phosphate
B	$(NH_4) NO_3$
C	Na-phosphate
D	$MgCl_2$
E	Malt extract
F	Na-citrate
G	Surfactant
H	Urea
I	$FeSO_4 \cdot 7H_2O$
J	$ZnSO_4 \cdot 7H_2O$
K	$MnSO_4 \cdot H_2O$
L	$CaCl_2 \cdot H_2O$

3. The Hadamard matrix (Table 3.11) is written for the variables given in Table 3.10.

Computation, regression analysis, and the decision of following Plackett-Burman design are discussed in Chapter 4.

TABLE 3.10

Variables with Their Levels

S. No.	Variables	+ Level at (kg/m³)	– Level at (kg/m³)
A	K-phosphate	2.50	2.00
B	$(NH_4)NO_3$	1.75	1.40
C	Na-phosphate	9.75	7.80
D	$MgCl_2$	0.037	0.03
E	Malt extract	12	9.60
F	Na-citrate	0.25	0.20
G	Surfactant	0.375	0.20
H	Urea	1.25	0.10
I	$FeSO_4·7H_2O$	0.00625	0.005
J	$ZnSO_4·7H_2O$	0.00129	0.001
K	$MnSO_4·H_2O$	0.00175	0.0014
L	$CaCl_2·H_2O$	0.0025	0.002
M, N, and O	Dummy variables	–	–

TABLE 3.11

Hadamard Matrix Generated by Cyclic Clockwise Arrangement of Rows

Run No.	A	B	C	D	E	F	G	H	I	J	K	L	(M)	(N)	(O)	Enzyme (U) as the Response
1	+	+	+	+	–	+	–	+	+	–	–	+	–	–	–	1.43
2	–	+	+	+	+	–	+	–	+	+	–	–	+	–	–	1.29
3	–	–	+	+	+	+	–	+	–	+	+	–	–	+	–	1.30
4	–	–	–	+	+	+	+	–	+	–	+	+	–	–	+	0.91
5	+	–	–	–	+	+	+	+	–	+	–	+	+	–	–	1.01
6	–	+	–	–	–	+	+	+	+	–	+	–	+	+	–	0.84
7	–	–	+	–	–	–	+	+	+	+	–	+	–	+	+	0.93
8	+	–	–	+	–	–	–	+	+	+	+	–	+	–	+	0.99
9	+	+	–	–	+	–	–	–	+	+	+	+	–	+	–	1.29
10	–	+	+	–	–	+	–	–	–	+	+	+	+	–	+	1.26
11	+	–	+	+	–	–	+	–	–	–	+	+	+	+	–	1.31
12	–	+	–	+	+	–	–	+	–	–	–	+	+	+	+	1.13
13	+	–	+	–	+	+	–	–	+	–	–	–	+	+	+	1.25
14	+	+	–	+	–	+	+	–	–	+	–	–	–	+	+	0.98
15	+	+	+	–	+	–	+	+	–	–	+	–	–	–	+	1.50
16	–	–	–	–	–	–	–	–	–	–	–	–	–	–	–	0.69

3.8 Second-Order Designs

Experimental designs from which the coefficient β_{ij} (in the model are non-zero) in addition to the coefficients of first-order terms $(\beta_1, ..., \beta_k)$ and quadratic-effect term, β_{ii}^2, can be estimated are called *second-order designs*.

β_i, β_{ij}, and β_{ii}^2 are coefficients of linear-effect, interaction-effect, and squared-effect terms, respectively, which are expressed in Equation 3.5, a second-order polynomial model.

$$\tilde{Y} = \beta_0 + \beta_i x_i + \beta_{ii}{}^2 x_i^2 + \beta_{ij} x_i x_j + \varepsilon \tag{3.5}$$

$\beta_{ij} x_i x_j$ is the interaction term, representing the effect between the ith and jth factors. This has additive effect. One can obtain such interaction in 2^k-designs. The coefficients of quadratic-effect terms (β_{ii}^2) cannot be estimated from 2^k-designs; however, this is possible in 3^k-designs. Therefore, 3^k-designs are second-order designs, whereas 2^k designs are first-order designs.[1]

In this section, the probable second-order designs are summarized, but both the Box-Behnken design and the CCD are analyzed in detail.

Before introducing different design plans, coding of variable is important.

3.8.1 Coding of Variables

Khuri and Cornell[6] proposed the distribution of three levels of input variables in the form of $(c_i - \delta_i)$, c_i, and $(c_i + \delta_i)$. The variable in coded form is $x_i = (X_i - c_i)/\delta_i$. X_i is the decoded value of the variable. δ_i is the distance from the centre value c_i.

In general,

$$x_i = \frac{2X_i - \left(X_{i\,\text{low}} + X_{i\,\text{high}}\right)}{X_{i\,\text{high}} - X_{i\,\text{low}}} \tag{3.6}$$

3.8.2 Koshal OFAT Second-Order Design

Koshal design of second order[3,4,7] contains at least three levels of explanatory variables. Specific observations of this design are as follows:

- Intuitive nature
- Saturated design
- Asymmetric around origin
- Design points with respect to linear, interaction, and quadratic effects
- Possibilities of improving the design, namely, exchange of design points for estimating quadratic effect and more spreading of interaction terms

The guidelines for interaction points in the design are as follows: Each coordinates has equal number of 1 and –1 levels for odd number of variables and for even number of variables. Coordinates for half of the explanatory variable will have one more 1 level than –1 level plus other half coordinates of explanatory variables of one more –1 than 1.

However, the design matrix appears to be of the following type for four variables of the second-order Koshal design.

$$D = \begin{vmatrix} 0 & 0 & 0 & 0 \\ 0 & 0 & 0 & 1 \\ 0 & 0 & 1 & 0 \\ 0 & 1 & 0 & 0 \\ 1 & 0 & 0 & 0 \\ -1 & 0 & 0 & 0 \\ 0 & -1 & 0 & 0 \\ 0 & 0 & -1 & 0 \\ 0 & 0 & 0 & -1 \\ 1 & 0 & 0 & 1 \\ 1 & 0 & 1 & 0 \\ 1 & 1 & 0 & 0 \\ 0 & 0 & 1 & 1 \\ 0 & 1 & 1 & 0 \\ 0 & 1 & 0 & 1 \end{vmatrix}$$

Ekman[4] defines the roatability criteria for such special designs.

3.8.3 3^k-Designs

In these designs, variable are at three levels: –1, 0, and +1 corresponding to low, intermediate, and high levels, respectively. Figure 3.21 shows the geometric form for two variables.

In this case, the regression model equation is Equation 3.7.

$$\hat{Y} = \beta_0 + \beta_1 x_1 + \beta_2 x_2 + \beta_{12} x_1 x_2 + \beta_{11} x_1^2 + \beta_{22} x_2^2 + \varepsilon \qquad (3.7)$$

In 3^k factorial designs, confounding, blocking, and fractional factorial are possible.[2]

3.8.4 Central-Composite Design

In 2^k or fractional 2^k designs, if centre points are added, the response surface explains the curvature. First-order design of 2^k is converted into second-order design. Those designs also contain another set of points called *axial points*. They are called *central-composite designs* (CCD). In the CCD,

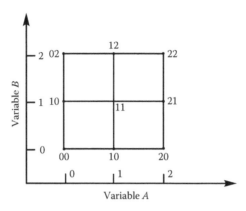

FIGURE 3.21
Geometric representations of 3^k-designs.

$$\text{Number of design points } (N) = \begin{bmatrix} \text{Factorial or fractional factorial} \\ \text{portion of design } (=F) \end{bmatrix}$$
$$+ \big[\text{number of axial points} (=2k) \big]$$
$$+ \begin{bmatrix} \text{Number of centre} \\ \text{points } (=n_c) \end{bmatrix}$$

Therefore, $N = F + 2k + n_c$ (3.8)

For k (= the number of variables) =2, in spherical CCD, $\alpha = \sqrt{2}$ and $n_c = 1$.
    Figure 3.22 represents the design points. A, B, C, and D are factorial points. E, F, G, and H are axial points (Figure 3.23).

$$\text{Design matrix} = \begin{vmatrix} 1 & 1 \\ -1 & 1 \\ 1 & -1 \\ -1 & -1 \\ \sqrt{2} & 0 \\ -\sqrt{2} & 0 \\ 0 & \sqrt{2} \\ 0 & -\sqrt{2} \\ 0 & 0 \end{vmatrix}$$

FIGURE 3.22
Design points.

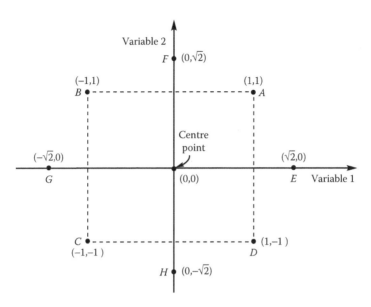

FIGURE 3.23
Design points of a CCD.

Figure 3.24 represents possible regions of the experiment (*cf.* Figure 3.22).

Each factor can have different levels and the position of the axial points from the centre points (= α) can be adjusted in the design. This gives different concepts of CCD. There are varieties of CCD: orthogonal, rotatable, orthogonal plus rotatable, spherical, and face-centred cube. According to Myers and Montgomery,[4] depending on the value of α, CCD can be face-centred cube, spherical, and rotatable.

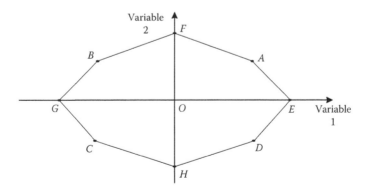

FIGURE 3.24
Possible experimental regions.

3.8.4.1 Types of CCD

1. For spherical CCD, $\alpha = \sqrt{2}$, for $k = 2$ and $n_c = 1$.
2. If $\alpha = 1$, it is face-centred cube CCD.
3. If $\alpha = \sqrt[4]{\text{number of factorial points}}$, CCD is rotatable. The conditions of second-order design being rotatable are as follows:

    a. All odd moments $= 0$, up to order four.

    b. All pure second-order moments are the same.

    c. Pure fourth-order moment, [iiii] $=$ three times mixed fourth-order moment ($=$[iijj])

    Pure second-order moment $= ([ii]) = F + 2\alpha^2$

    Pure fourth-order moment $= [iiii] = F + 2\alpha^4$

    Mixed fourth-order even moment $= [iijj] = F$

    From the third condition $= F + 2\alpha^4 = 3F$

$$\alpha = \sqrt[4]{F}$$

4. For orthogonal CCD, mixed fourth-order moment [iijj] $= 1^6$
5. For orthogonal as well as rotatable CCD, α and n_c both need to satisfy certain values. General formula for a CCD is both orthogonal and rotatable.

$$n_c = 4\sqrt{F} + 4 - 2k \text{ and } \alpha^4 = F$$

Example 3.5

Recall Example 3.1. Apply the concept of CCD to calculate the value of α and number of centre points for CCD being both orthogonal and rotatable.
For two variables, that is, temperature and pH of fermentation broth

$$4\sqrt{4} + 4 = 2k + n_c$$

$$8 + 4 = 2 \times 2 + n_c$$

$$n_c = 8$$

and

$$\alpha = \sqrt[4]{4}$$

Example 3.6

Write the matrix for CCD in a general biological system.

Solution:
If the experimenter does not consider any spherical cases of CCD, the matrix is based on the general equation for total number (N) of experimental runs in a CCD as per the following equation.

$$N = F + 2k + n_c$$

TABLE 3.12

Matrix of a CCD in General

	Variable 1	Variable 2	Portion of the Design Matrix
$D =$	-1	-1	
	-1	$+1$	
	$+1$	-1	Factorial portion
	$+1$	$+1$	
	0	$-\alpha$	
	0	$+\alpha$	
	$-\alpha$	0	Axial portion
	$+\alpha$	0	
	0	0	←Centre point

or

$$N = 2^k + 2k + n_c$$

For two variables, $k = 2$, variable 1 = aeration rate, and variable 2 = agitation rate, which should be maintained in a bioreactor for the synthesis of penicillin acylase.

1. Total number of experimental points in full factorial portion of the design = $2^2 = 4$
2. Total number of experimental points in the axial portion of the design = $2 \times 2 = 4$
3. Number of replicates in the centre point = $n_c = 1$

The matrix is given in Table 3.12.

3.8.4.2 Coding of Variables

Khuri and Cornell[6] proposed the distribution of three levels of input variables. This is described in Section 3.8.1.

3.8.4.3 Matrix for Special Cases of CCD

1. Table 3.13 describes the matrix for spherical CCD. The condition is $\alpha = \sqrt{2}$.
2. Table 3.14 is the matrix for face-centred cube, where $\alpha = 1$.
3. For rotatable CCD with two variables, $\alpha = \sqrt[4]{F} = \sqrt[4]{4}$. Axial points in the matrix will be replaced by $\sqrt[4]{4}$.
4. For an orthogonal plus rotatable CCD, similarly design matrix is established to run the experiments as per conditions calculated in the Example 3.5.

TABLE 3.13

Matrix for Spherical CCD

$$D = \begin{bmatrix} -1 & -1 \\ -1 & +1 \\ +1 & -1 \\ +1 & +1 \\ 0 & -\sqrt{2} \\ 0 & +\sqrt{2} \\ -1 & 0 \\ +\sqrt{2} & 0 \\ 0 & 0 \end{bmatrix}$$

TABLE 3.14

Matrix for Face-Centred Cube

$$D = \begin{bmatrix} -1 & -1 \\ -1 & +1 \\ +1 & -1 \\ +1 & +1 \\ 0 & -1 \\ 0 & +1 \\ -1 & 0 \\ +1 & 0 \\ 0 & 0 \end{bmatrix}$$

Example 3.7

This example concerns with the coding of experimental variables followed by the design matrix as per CCD for enzyme production.

Solution:

Coding of variables: Phosphate, ammonium nitrate, malt extract, surfactant, and minerals influence the enzyme production. These variables require further analysis in the second-order design.

Table 3.15 shows variables in a CCD for optimization of enzyme production by a fungus. X_i is the actual values of the experimental variables. x_i is the corresponding coded values of the variable.

The design matrix is shown in Table 3.16.

$$\begin{aligned} \text{Number of experimental runs} = N &= 2^{5-1} \left(= \text{Fractional factorial points} \right) \\ &+ 2 \times 5 \ \left(= \text{Number of axial points} \right) \\ &+ n_c \left(= \text{Number of centre points} \right) \end{aligned}$$

TABLE 3.15

Coded and Actual Values of Variables

	Coded Values (x_i)				
	$-\alpha$ (−2)	−1	0	+1	$+\alpha$(+2)
Variables	Actual Values (kg/m³)				
Phosphate (X_1)	0	3.841	7.682	11.523	15.364
Ammonium nitrate (X_2)	0	0.875	1.750	2.625	3.500
Malt extract (X_3)	0	0.625	1.25	1.875	2.500
Surfactant (X_4)	0	0.1250	0.250	0.375	0.500
Minerals (X_5)	0	0.0196	0.0393	0.059	0.0787

TABLE 3.16

Design Matrix in the Coded Form of the Variables

Run No.	Phosphate (x_1)	Ammonium Nitrate (x_2)	Malt Extract (x_3)	Surfactant (x_4)	Minerals (x_5)
1	−1	−1	−1	−1	1
2	1	−1	−1	−1	−1
3	−1	1	−1	−1	−1
4	1	1	−1	−1	1
5	−1	−1	1	−1	−1
6	1	−1	1	−1	1
7	−1	1	1	−1	1
8	1	1	1	−1	−1
9	−1	−1	−1	1	−1
10	1	−1	−1	1	1
11	−1	1	−1	1	1
12	1	1	−1	1	−1
13	−1	−1	1	1	1
14	1	−1	1	1	−1
15	−1	1	1	1	−1
16	1	1	1	1	1
17	−2	0	0	0	0
18	2	0	0	0	0
19	0	−2	0	0	0
20	0	2	0	0	0
21	0	0	−2	0	0
22	0	0	2	0	0
23	0	0	0	−2	0
24	0	0	0	2	0
25	0	0	0	0	−2
26	0	0	0	0	2

(*Continued*)

TABLE 3.16 (*Continued*)

Design Matrix in the Coded Form of the Variables

Run No.	Phosphate (x_1)	Ammonium Nitrate (x_2)	Malt Extract (x_3)	Surfactant (x_4)	Minerals (x_5)
27	0	0	0	0	0
28	0	0	0	0	0
29	0	0	0	0	0
30	0	0	0	0	0
31	0	0	0	0	0
32	0	0	0	0	0

Experiments are carried out in each run conditions at least three times. Responses for individual experiments are recorded, which will be analyzed as per the procedures in Chapter 4. In this case, other experimental conditions are volume of the enzyme production medium − 100 cm³ in 500 cm³ Erlemeyer flask, age of the organism − 120 h, inoculums age − 36 h, inoculum level − 0.24 kg/m³ dry equivalent of cells, approximately, fermentation time − 216 h, shaker speed − 180 rpm, temperature of fermentation − 30°C, initial pH of the production medium − 5. The values for the concentrations of medium constituents, which are variables in these experiments, are given in Table 3.15. The rest of the medium compounds (kg/m³) are complex C-source 10; Na citrate, 9.6, urea, 0.3; $ZnSO_4 \cdot H_2O$, 0.0014.

3.8.5 Box-Behnken Design

The Box-Behnken design is an alternative to CCD. This is a second-order rotatable or nearly rotatable three-level incomplete factorial designs.[8]

3.8.5.1 Characteristics of the Design Points

1. This is applicable for at least three variables.
2. In the design matrix, two variables at a time will appear with full factorial points.
3. Rest of the variables is at 0-level.
4. Design points lie on the surface of the geometric figure centred at the origin and tangential to the mid-point of each edge of the geometric figure.
5. At least one point is in the 0-level.
6. There is no design point at either higher or lower levels.
7. Orthogonal blocking is desirable.
8. Second-order model is applicable to find optimal conditions.

3.8.5.2 Differences between the CCD and the Box-Behnken Design

Table 3.17 gives the summary of differences in the design matrices.

3.8.5.3 Design Matrix

Table 3.18 describes the design matrix for four variables, namely, pH (controlled, x_1), temperature (x_2), agitation (x_3), and aeration rate (x_4) in a fermentation

TABLE 3.17

Characteristic Differences between the CCD and the Box-Behnken Design

S. No.	CCD	Box-Behnken Design
1.	Two-level factorial or fractional factorial with axial and centre points appear in the CCD.	Variables at three levels and hence three-level incomplete factorial rotatable designs in blocks, having centre points are in the matrix.
2.	Design points have all variables at higher or lower level.	No design points have variables at higher or lower level.
3.	Designs are either rotatable, orthogonal, spherical, face-centred cube, or rotatable plus orthogonal.	This is a block orthogonal design.

TABLE 3.18

Matrix for the Box-Behnken Experimental Design

	Variable 1 (x_1)	Variable 2 (x_2)	Variable 3 (x_3)	Variable 4 (x_4)	Set Number with the Blocks
$D =$	+1	+1	0	0	Set –1 consists
	–1	–1	0	0	of blocks 1
	–1	+1	0	0	and 2
	+1	–1	0	0	
	0	0	+1	+1	
	0	0	–1	–1	
	0	0	–1	+1	
	0	0	+1	–1	
	0	0	0	0	
	+1	0	+1	0	Set –2 consists
	–1	0	–1	0	of blocks 3
	–1	0	+1	0	and 4
	+1	0	–1	0	
	0	+1	0	+1	
	0	–1	0	–1	
	0	–1	0	+1	
	0	+1	0	–1	
	0	0	0	0	

(Continued)

TABLE 3.18 (*Continued*)

Matrix for Box-Behnken Experimental Design

Variable 1 (x_1)	Variable 2 (x_2)	Variable 3 (x_3)	Variable 4 (x_4)	Set Number with the Blocks
+1	0	0	+1	Set −3 consists
−1	0	0	−1	of blocks 5
−1	0	0	+1	and 6
+1	0	0	−1	
0	+1	+1	0	
0	−1	−1	0	
0	+1	−1	0	
0	−1	+1	0	
0	0	0	0	

experiment. Coding of the variables is as per the method described in Section 3.8.1. The matrix elements are as per the characteristics given in Box-Behnken design.

3.8.6 Doehlert Designs

In this design, different variables are at different numbers of levels.[9] This is generated from a regular simplex containing $(k + 1)$ points, where k is the number of variables. The conditions for first few runs are the coordinates of the vertices of the simplex. The subsequent runs are generated by

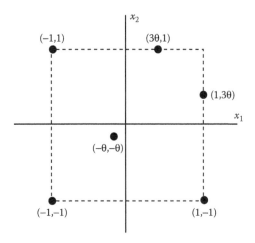

FIGURE 3.25

Saturated designs of Box and Draper for two experimental variables.

subtracting every run in the simplex from each other. The simplex moves around the design centre. These designs are neither orthogonal nor rotatable. Non-uniform variance of the predicted values is observed over the experimental range. To cover the experimental range, this design is better.

3.8.7 Some Special Saturated or Near Saturated Designs

Roquemore[10] suggested in a same design a few variables in the CCD format and other variables in different levels, so that odd moments are zero and all second pure moments are equal with a view to near rotatable design. Hoke design,[11] Box and Draper design[12] (*cf.* Figure 3.23), and Notz design[13] are important in this context. Notz design matrix contains a mixture of design points of OFAT and partial 2^k-design points. On the other hand, Hoke design is an irregular fractions of 3^k-designs. There are computer-generated designs as well.

Exercises

3.1 Explain the response surface methodology. Consider a suitable example of a bioprocess.

3.2 What are the differences between a first-order experimental design and a second-order experimental design?

3.3 What are the differences between the Box-Behnken design and the 3^k full factorial experimental design plan?

3.4 State the criteria for the Plackett-Burman design. Sucrose, yeast extract, KH_2PO_4, and $NaNO_3$ are medium constituents for enzyme production. They are the variables. Code them properly and write the Hadamard matrix.

3.5 The following factors need optimization statistically for the bioprocess development: temperature, pH, and agitation. Apply the Box-Behnken design and the CCD in this regard. Will you achieve the same optima by these two experimental designs? Explain.

3.6 For β-1,3 glucanase production by *Trichoderma harzianum* in a bioreactor pH (controlled), agitation rate, and aeration rate were the variables. Centre point values were 5, 160 rev/min, and 1.5 (m^3 air)/(m^3 medium) minute for pH (controlled), agitation rate, and aeration rate, respectively. How will you formulate the design matrix, if you consider the Box-Behnken design procedure?

3.7 Table 3.7.1 summarizes an experimental plan. Name it. Justify the same by mentioning the properties of the design.

TABLE 3.7.1

Variables at Two Levels with the Responses

Run #	A	B	C	D	E	F	G	H	I	J	K	Metabolite (kg/m³)
					Levels/Variables							
1	+	+	−	+	+	+	−	−	−	+	−	0.41
2	−	+	+	−	+	+	+	−	−	−	+	0.51
3	+	−	+	+	−	+	+	+	−	−	−	0.38
4	−	+	−	+	+	−	+	+	+	−	−	0.47
5	−	−	+	−	+	+	−	+	+	+	−	0.47
6	−	−	−	+	−	+	+	−	+	+	+	0.30
7	+	−	−	−	+	−	+	+	−	+	+	0.31
8	+	+	−	−	−	+	−	+	+	−	+	0.51
9	+	+	+	−	−	−	+	−	+	+	−	0.49
10	−	+	+	+	−	−	−	+	−	+	+	0.51
11	+	−	+	+	+	−	−	−	+	−	+	0.44
12	−	−	−	−	−	−	−	−	−	−	−	0.39

3.8 Initial pH of enzyme production medium and temperature on the production of extracellular enzyme is given in Table 3.8.1. Write the coded values of the variables. Give the analysis of this 2^2 design.

3.9 Name the design plan. Code the variables as per the design requirement. Rewrite the design matrix using coded values for the data in Table 3.9.1.

3.10 For proper production of a metabolite, biological parameters are important to control. Table 3.10.1 gives such information. Code the experimental variables properly and write the design matrix in coded form. What is the value of α? Classify the design plan as per the value of α.

TABLE 3.8.1

Decoded Values and Response for Analysis Using 2^2 Design

Run No.	Initial pH X_1	Temperature (°C) X_2	Chitinase (U) Experimental (Y)	
			Set 1	Set 2
1	5	25	0.1235	0.0890
2	7	25	0.0301	0.0337
3	5	35	0	0
4	7	35	0	0
5	6	30	0.1542	0.1630

TABLE 3.9.1

Decoded Values of Initial pH and Temperature

Run No.	Initial pH X_1	Temperature (°C) X_2
1	5	25
2	7	25
3	5	35
4	7	35
5	6	30
6	6	30
7	6	30
8	4.59	30
9	7.41	30
10	6	23
11	6	37
12	6	30
13	6	30
14	6	30

TABLE 3.10.1

Conditions of Age of Organism and the Age of Inoculum

Run No.	Age of the Organism (h) X_1	Inoculum Age (h) X_2
1	48	24
2	96	24
3	48	48
4	96	48
5	72	36
6	72	36
7	72	36
8	38	36
9	106	36
10	72	19
11	72	53
12	72	36
13	72	36
14	72	36

3.11 Is Table 3.11.1 a CCD? If so, which class of CCD? How will you convert this design into the Box-Behnken design format? Make suitable modification of the given plan to construct the design plan as per Box-Behnken.

3.12 An experimenter made the design plan as per Table 3.12.1. Which design will conform to this specification of experiments? If you need

TABLE 3.11.1

Variables Affecting the Number of Heterokayons Production Using Electroporation

Run No.	Pulse Voltage, volts/mm (X_1)	Pulse Width, µs (X_2)	No. Pulses
1	400	75	7
2	450	50	6
3	450	50	6
4	400	25	5
5	500	75	5
6	500	25	7
7	500	25	5
8	500	75	7
9	450	50	6
10	400	25	7
11	400	75	5
12	450	50	6
13	450	7.95	6
14	450	50	4.318
15	534.10	50	6
16	450	50	7.682
17	450	50	6
18	450	50	6
19	365	50	6
20	450	92.05	6

TABLE 3.12.1

Influencing Factors for Organic Acid Fermentation in Batch Bioreactor

Run No.	pH	Aeration (l)/(l)(min)	Agitation (rev/min)
1	6	1	160
2	7	1	160
3	6.5	1.5	160
4	7	1.5	100
5	6.5	2	220
6	6	1.5	100
7	6	1.5	220
8	6.5	2	100
9	7	1.5	220
10	6.5	1	220
11	7	2	160
12	6.5	1.5	160
13	6.5	1	100
14	6.5	1.5	160
15	6	2	160

to run as per the following plan, what modification do you need to introduce to the design plan? Suggested plans are as follows:

a. Koshal second-order design

b. CCD with orthogonal plus rotatable design

c. Box-Behnken design

References

1. Johnson NL and Leone FC (Eds.), *Statistics and Experimental Design in Engineering and Physical Sciences*, Vol. II, John Wiley & Sons, New York, 1977.

2. Montgomery DC (Ed.), *Design and Analysis of Experiments*, 7th edition, Wiley, India, 2009.

3. Koshal RS, Application of the method of maximum likelihood to the improvement of curves fitted by the method of moments. *Journal of Royal Statistical Society*, **A96**, 303–313, 1933.

4. Ekman C, Saturated designs for second order models, Research Report 1994: 9, Department of Statistics, Gothenburg University, Sweden, 1994.

5. Plackett RL and Burman JP, The design of optimum multifactorial experiments, *Biometrika*, **33**, 305–325, 1946.

6. Khuri AI and Cornell JA (Eds.), Response surfaces designs and analyses, Marcel Dekker, New York, 1987.

7. Meyers RH and Montgomery DC (Eds.), Response surface methodology: *Process and Product Optimization Using Designed Experiments*, 2nd edition, Wiley, New York, 2002.

8. Box GEP and Behnken DW, Some new three level designs for the study of quantitative variables, *Techometrics*, **2**, 455–476, 1960

9. Doehlert DH, Uniform shell designs, *Journal of Royal Statistical Society*, **C19**, 231–239, 1970.

10. Roquemore KG, Hybrid designs for quadratic response surfaces, *Techometrics*, **18**, 419–423, 1976.

11. Hoke AT, Economical second-order designs based on irregular fractions of the 3^n factorial, *Techometrics*, **16**, 375–384, 1974.

12. Box GEP and Draper NR, Factorial designs, the $|X'X|$ criterion and some related matters, *Techometrics*, **13**, 731–741, 1971.

13. Notz W, Saturated designs for multivariate cubic regression, Mimeograph series/ Dept. of Statistics, Division of Mathematical Sciences, Purdue University, West Lafayette, IN, 1979.

4

Statistical Analysis of Experimental Designs and Optimization of Process Variables

OBJECTIVE: The statistical analysis of experimental design consists of creating a model from the data and testing the model assumptions. The results of the analysis are used to determine the important factors and optimum settings.

4.1 Introduction

In Chapter 3, experimental designs either of first order or of second order have been discussed elaborately with respect to formulation of the experimental plan and the design matrix. As per the experimental plan, it is necessary to carry out the experiments to get the suitable response and/or responses. Biological processes are good examples of a multi-response system. The relation among input experimental variables on the response is necessary to be established for proper functioning of the biological process. Not all factors influence the response in a similar fashion. So, the selection of proper experimental design, role of influencing factors for the system, the relation of influencing factors with the response from the experiments, and the optimal or near-optimal conditions for the input factors are the part of analysis of experimental designs. This chapter will discuss the overall analysis of different class of experimental designs. Chapters 5 and 6 emphasize on a few special designs, namely, evolutionary operation programme (EVOP) and Taguchi's design. In this chapter, initial discussion refers to the estimation of statistical parameters, pertaining to the experimental designs, followed by finding optimal conditions of experimental variables for single-response and multi-response systems, using response surface methodology. Modelling and displacement are the two stages in response surface methodology, which will direct to optimal conditions of the process.[1] Model will be simple and will correlate experimental variables with the response of the process.

4.2 Analysis of Experimental Designs

Organization of this section is the analyses of the following experimental design.

- First-order designs are as follows:
  - One-factor-at-a-time (OFAT) design of Koshal
  - The simplex design
  - 2^k-design
  - Screening design of Plackett-Burman
- Second-order designs include the following:
  - OFAT design of Koshal
  - 3^k-design
  - Central-composite design (CCD)
  - Box and Behenken design.

The model and allied information
The theoretical concept for different order designs is the requirement of this section.

1. *For the first-order designs*

   For 2^2-design centered at (0,0) is the example of first-order design. The design matrix for this case is given in Table 4.1. The first option of Figure 3.12 is the geometric representation of the design points.

   The model equation is

   $$\hat{Y} = \beta_0 + \beta_1 x_1 + \beta_2 x_2 \tag{4.1}$$

TABLE 4.1

2^2-Design Matrix Centered at (0,0)

Run Number	pH	Temperature	Response, Gluconic Acid (kg/m³)
1	−1	−1	P_1
2	+1	−1	P_2
3	−1	+1	P_3
4	+1	+1	P_4
5	0	0	P_5
6	0	0	P_6
7	0	0	P_7

The design matrix

$$
X = \begin{bmatrix}
1 & -1 & -1 \\
1 & +1 & -1 \\
1 & -1 & +1 \\
1 & +1 & +1 \\
1 & 0 & 0 \\
1 & 0 & 0 \\
1 & 0 & 0
\end{bmatrix}
$$

$\uparrow$ $\uparrow$

β_0 (coded values for the factor levels)

The response matrix

$$
Y = \begin{bmatrix}
P_1 \\
P_2 \\
P_3 \\
P_4 \\
P_5 \\
P_6 \\
P_7
\end{bmatrix}
$$

Following the usual procedure

$$
X'X = \begin{bmatrix} \quad \end{bmatrix}
$$

and

$$
X'Y = \begin{bmatrix} \quad \end{bmatrix}
$$

are calculated. Then, using

$$\beta = (X'X)^{-1}X'Y \tag{4.2}$$

From the triplicate runs at the centre point estimate of the variance of response values σ^2 is calculated. Then substitution of this value in

$$V(\beta) = (X'X)^{-1}\sigma^2 \tag{4.3}$$

Taking the square roots of the elements on the main diagonal, the standard error of β_0 and β_1 are evaluated. Equation 4.4 is the complete equation.

$$\hat{Y} = \beta_0 \pm \beta_1 x_1 \pm \beta_2 x_2 \tag{4.4}$$

Analysis of variance (ANOVA) for fitting the model equation contains the terms, which are calculated for specific example (see Table 4.2).

a. *Conclusion from ANOVA*: Residuals should be small. If ideal model matches exactly with the observed response, there will be no residuals. ANOVA is the quantitative representation of quality of fit of a model.

$$\text{Total sum squares (SS)} = \sum \left(\hat{Y}_i - \bar{Y} \right)^2 + \sum \left(Y_i - \hat{Y}_i \right)^2$$

$$= \text{Regression SS} + \text{Residual SS} \tag{4.5}$$

$$\text{SST} = \text{SSR} + \text{SSE}$$

Significance: Larger fraction contributed by regression suggests that the model is better.

Another statistic R^2 is called *coefficient of determination* of the model, which is calculated by Equation 4.6.

TABLE 4.2

Elements of ANOVA

Source of Variation	Sum Square	Degrees of Freedom	Mean Square
Regression			
Residual			
Lack of fit			
Pure error			
Total			

$$R^2 = \frac{SSR}{SST} = \frac{\sum(\hat{Y}_i - \bar{Y})^2}{\sum(Y_i - \bar{Y})^2} \times 100\,\% \tag{4.6}$$

If $R^2 = 100\%$, there is no residuals. All the variation about the average is explained by the regression equation, that is, the model will fit the observed values. $\sum(Y_i - \bar{Y}) = 0$, which, uses 1 degree of freedom, that is, $(n-1)$ degrees of freedom. SST is associated with $(n-1)$ degrees of freedom. The degree of freedom associated with SSE is $(n-p)$. For SSR, $(p-1)$ is the degree of freedom (see Table 4.3). p is the number of parameters in the fitted model.

b. Lack-of-fit and pure error: Distribution of residuals indicates the correctness of the model. This may suggest for any improvement necessary for the model.

Bruns et al.[1] suggested the relations using the following equations:

$$\text{SSE at the } i\text{th level} = (SSE)_i = \sum_j^{n_i}\left(Y_{ij} - \hat{Y}_i\right)^2$$

SSE = Pure error SS associated with $(n-m)\text{df}$ + Lack of fit SS, associated with $\text{df} = (m-p)$ \hfill (4.7)

where:
   n is the number of observations
   m is the number of levels of variable

$$MS_{\text{Pure error}} = \frac{\sum_i^m \sum_j^{n_j}\left(Y_{ij} - \bar{Y}_i\right)^2}{(n-m)} \tag{4.8}$$

TABLE 4.3

ANOVA Table

Sources of Variation	Sum Square	Degrees of Freedom	Mean Square
Regression	$SSR = \sum\left(\hat{Y}_i - \bar{Y}\right)^2$	$(p-1)$	$\dfrac{SSR}{(p-1)} = MSR$
Residual	$SSE = \sum\left(Y_i - \hat{Y}_i\right)^2$	$(n-p)$	$\dfrac{SSE}{(n-p)} = MSE$
Total	$SS = \sum\left(Y_i - \bar{Y}\right)^2$	$(n-1)$	

$$MS_{\text{Lack of fit}} = \frac{\sum\limits_{i}^{m}\sum\limits_{j}^{n_j}\left(\hat{Y}_i - \bar{Y}_i\right)^2}{(m-p)} \tag{4.9}$$

Maximum percent variation

$$\text{explained by the model} = \left(\frac{\text{SST} - \text{SSPE}}{\text{SST}}\right) \times 100\ \%$$

2. *For the second-order designs*

The following is the design matrix for two variables:

$$
D = \begin{array}{cc}
x_1 & x_2 \\
\begin{bmatrix}
-1 & -1 \\
+1 & -1 \\
-1 & +1 \\
+1 & +1 \\
-1.414 & 0 \\
+1.414 & 0 \\
0 & -1.414 \\
0 & +1.414 \\
0 & 0 \\
0 & 0 \\
0 & 0 \\
0 & 0
\end{bmatrix}
\end{array}
$$

The second-order model is

$$Y = \beta_0 + \beta_1 x_1 + \beta_2 x_2 + \beta_{11}x_1^2 + \beta_{22}x_2^2 + \beta_{12}x_1 x_2 + \varepsilon \tag{4.10}$$

The matrix X and vector $\vec{Y}$ for this model

$$
X = \begin{array}{cccccc}
& x_1 & x_2 & x_1^2 & x_2^2 & x_1 x_2 \\
\begin{bmatrix}
1 & -1 & -1 & 1 & 1 & 1 \\
1 & 1 & -1 & 1 & 1 & -1 \\
1 & -1 & 1 & 1 & 1 & -1 \\
1 & 1 & 1 & 1 & 1 & 1 \\
1 & -1.414 & 0 & 2 & 0 & 0 \\
1 & 1.414 & 0 & 2 & 0 & 0 \\
1 & 0 & -1.414 & 0 & 2 & 0 \\
1 & 0 & 1.414 & 0 & 2 & 0 \\
1 & 0 & 0 & 0 & 0 & 0 \\
1 & 0 & 0 & 0 & 0 & 0 \\
1 & 0 & 0 & 0 & 0 & 0 \\
1 & 0 & 0 & 0 & 0 & 0 \\
\end{bmatrix}
\end{array}
\quad
\vec{Y} = \begin{bmatrix}
y_1 \\ y_2 \\ y_3 \\ y_4 \\ y_5 \\ y_6 \\ y_7 \\ y_8 \\ y_9 \\ y_{10} \\ y_{11} \\ y_{12}
\end{bmatrix}
$$

The matrix

$$
X'X = \begin{bmatrix}
12 & 0 & 0 & 8 & 8 & 0 \\
0 & 8 & 0 & 0 & 0 & 0 \\
0 & 0 & 8 & 0 & 0 & 0 \\
8 & 0 & 0 & 12 & 4 & 0 \\
8 & 0 & 0 & 4 & 12 & 0 \\
0 & 0 & 0 & 0 & 0 & 4
\end{bmatrix}
$$

$$
\text{Vector } X'\vec{Y} = \begin{bmatrix} \\ \\ \\ \\ \\ \\ \end{bmatrix}
$$

$$b = (X'X)^{-1}X'\vec{Y} = \begin{bmatrix} \beta_0 \\ \beta_1 \\ \beta_2 \\ \beta_{11} \\ \beta_{22} \\ \beta_{12} \end{bmatrix}$$

Table 4.4 is the ANOVA table for this example.

Other statistic are as follows:

$$(a) \; R^2 = \frac{SSR}{SST} = \left(1 - \frac{SSE}{SST}\right),$$

$$R_{adjusted}^2 = 1 - \frac{SSE/(n-p)}{SST/(n-1)} = 1 - \frac{(n-1)}{(n-p)}(1 - R^2)$$

$$(b) \; PRESS \; = \sum_{i=1}^{n}\left(\frac{e_i}{1-h_{ii}}\right)^2$$

$$R_{prediction}^2 = 1 - \frac{PRESS}{SST}$$

(4.11)

h_{ii} is the diagonal element of hat matrix.[2]

a. *Significance*

R^2 is a measure of the reduction in the variability of response using $x_1, x_2, x_3, \ldots, x_x$ in the model. The range of R^2 is $0 \le R^2 \le 1$. R^2 does not necessarily imply the regression model is a good one.

TABLE 4.4

ANOVA

Sources of Variation	Sum Squares	Degrees of Freedom	Mean Square	F	P-Value
Regression		5			
a. $SSR_1(\beta_1, \beta_2 \vert \beta_0)$	( )	(2)			
b. $SSR_2(\beta_{11}, \beta_{22}, \beta_{12} \vert \beta_0, \beta_1, \beta_2)$	( )	(3)			
Residual					
Lack of fit	( )				
Pure error	( )				
Total					

R² *adjusted* differs from R² value. Sometimes this will suggest that additional term has little effect on R²–adjusted.

R – student stands for externally standardized residual.

$$t_i = \frac{e_i}{\sqrt{(S_{(i)}^2(1-h_{ii})}}; \quad i = 1, 2, 3, ..n \tag{4.12}$$

$$S_{(i)}^2 = \frac{(n-p)MSE - [e_i/(1-h_{ii})]}{(n-p-1)} \tag{4.13}$$

Test statistic for lack of fit (lof)

$$F_{lof} = \frac{[SSLOF/(m-p)]}{[SSPE/(n-m)]} = \frac{MSLOF}{MSPE} \tag{4.14}$$

Cases:

1. If the regression function is linear, F_{lof} follows $F_{(m-p),(n-m)}$ distribution.
2. If $F_{lof} > F_{(m-p),(n-m)}$, the regression function is not linear.

4.2.1 First-Order Designs

4.2.1.1 OFAT Design of Koshal

The matrix for Koshal design is mentioned in Chapter 3 for first-order designs with and without interaction of variables.[3]

Example 4.1

Pulse width, field strength, and electroporation buffers are the important variables under investigation to achieve higher transformants/μg DNA. Three variables are coded in Table 4.5.

The design matrix is Table 4.6.

Four coefficients are estimated in the model (Equation 4.15). The calculation is given in Appendix A.2.

$$\tilde{Y} = \beta_0 + \beta_1 x_1 + \beta_2 x_2 + \beta_3 x_3 + \varepsilon \tag{4.15}$$

TABLE 4.5

Coding of Variables for Koshal Design

Field Strength (kV/cm)		Pulse Width (ms)		Electroporation Buffer	
3	3.5	3	15	1M sorbitol	(1M sorbitol + 20 mM HEPES)
0 level	1 level	0 level	1 level	0 level	1 level

TABLE 4.6

OFAT Design of Koshal with No Interaction of Variables

Run Number	x_1	x_2	x_3	Response, Transformants/ μg DNA
1	0	0	0	0.89×10^5
2	1	0	0	10^4
3	0	1	0	10^3
4	0	0	1	2.4×10^6

4.2.1.2 Simplex Design

This is an orthogonal first-order design, following the OFAT procedure.[4] The design points represent the vertices of a regular geometric configuration (Figures 4.1 and 4.2). Number of design points is equal to the number of variables plus one. A detailed procedure is described by Khuri and Cornell.[4]

4.2.1.3 2^k-Design

Chapter 3 gives the details of this design with analysis.

4.2.1.4 Screening Design of Plackett-Burman

Example 4.2

Let *A, B, C, D, E, F, G, H, I, J, K, L*, and *M* be the various chemical variables. The following levels for each variable are assumed in this case (*cf.* Table 4.7).

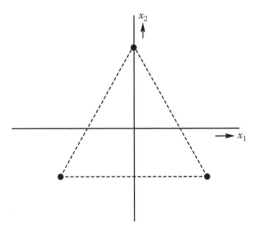

FIGURE 4.1
Simplex concept for two variables.

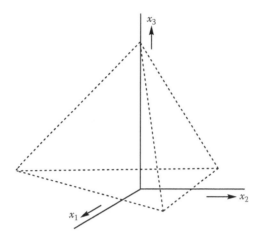

FIGURE 4.2
Simplex concepts for three variables.

TABLE 4.7

Levels of the Variables

Variables	Maximum (Coding: +1 or +)	Minimum (Coding: −1 or −)
A	1.75	1.4
B	2.5	2
C	11.11	8.89
D	0.037	0.0296
E	12	9.6
F	1.25	1
G	0.375	0.3
H	0.25	0.2
I	5.683	4.546
J	1.255×10^{-3}	1.004×10^{-3}
K	0.65×10^{-3}	0.52×10^{-3}
L	0.569×10^{-3}	0.455×10^{-3}
M	0.681×10^{-3}	0.54×10^{-3}

Note: N and O are dummy variables.

4.2.1.4.1 *Formulation of Design Matrix*

Since the number of the variables given are 13. In order to form the design matrix as per Plackett and Burman design,[5] the design requires two dummy variables (e.g. N and O). So, the number of runs = $(k + 1) = 16$ (including the two dummy variables). The concept of design matrix is given in Section 3.7.1.

Rules of forming the design matrix is mentioned in the Chapter 3, under Section 3.7.1.1.

As per experiment, the response is given in the design matrix (Table 4.8).

TABLE 4.8

Response Data Corresponding to
the Designed Experimental Runs
at a Particular Time of the Process

Variables/Runs	Responses
1	0.4816
2	0.3750
3	0.4823
4	0.5105
5	0.4203
6	0.3647
7	0.3933
8	0.4859
9	0.6515
10	0.4420
11	0.4218
12	0.3858
13	0.3840
14	0.3803
15	0.3657
16	0.3760

Calculation of effect of experimental variables

$$\text{Effect} = \left[\frac{\text{Response at} (+) \text{ level}}{8} \right] - \left[\frac{\text{Response at} (-) \text{ level}}{8} \right]$$

For example,

Effect of variable E

$$= \left(\frac{0.375 + 0.5105 + 0.4203 + 0.4859 + 0.3858 + 0.384 + 0.3803 + 0.3657}{8} \right)$$

$$- \left(\frac{0.4816 + 0.4823 + 0.3647 + 0.3933 + 0.6515 + 0.442 + 0.4218 + 0.376}{8} \right)$$

$$= 0.4134375 - 0.45165$$

$$= -0.0382125$$

TABLE 4.9

Effects of the Variables

Variables	Effects
A	0.0692375
B	-0.0086125
C	-0.0368125
D	0.017513
E	-0.03821
F	-0.00519
G	-0.05976
H	0.024488
I	0.070663
J	0.031613
K	-0.00096
L	-0.02569
M	-0.013313
N	0.011863
O	0.049638

Effects of other variables are tabulated in Table 4.9.

$$\text{Mean effect} = \frac{\Sigma \text{ Response}}{\text{Number of runs}} = \frac{6.9207}{16} = 0.432544$$

- Sample size = 16
- Sample average = $(\Sigma x_u)/N = 0.432544$
- Sample variance = $\text{var}(x) = \Sigma(x_u - x_{av})^2/(N-1) = 0.005766211$
- Standard deviation (SD) = $(\sqrt{\text{var}(x)}) = 0.075935569$
- Standard error (SE) = $(\sqrt{\text{var}(x)/n}) = (\text{SD})/\sqrt{n} = 0.018983892$

Calculation of t-value

$$t(x_i) = \frac{E(x_i)}{\text{SE}} = \frac{\text{Effect of variable}}{\text{Standard error}}$$

Calculation of *t*-value for variable $E = (-0.03821/0.018983892) = -2.012759003$
Other *t*-values are shown in Table 4.10.

Probability due to chance
When $t_{cal} > t_{tab}$, probability due to chance will be almost equal to *t*-distribution value at $\alpha = 0.025$ (two-tail) with degree of freedom $(N-1=15)$ (*cf.* Table 4.11).

TABLE 4.10

Variables and Corresponding *t*-Values

Variables	Effects	*t*-Value
A	0.0692375	3.64719275
B	-0.0086125	-0.453542398
C	-0.0368125	-1.939143963
D	0.017513	0.92249295
E	-0.03821	-2.012759003
F	-0.00519	-0.273389668
G	-0.05976	-3.147931904
H	0.024488	1.289935684
I	0.070663	3.722260912
J	0.031613	1.66525387
K	-0.00096	-0.050569187
L	-0.02569	-1.35325252
M	-0.013313	0.701278739
N	0.011863	0.624871858
O	0.049638	2.614716698

TABLE 4.11

Values of Probability Due to Chance

Variables	*t*-Value	|*t*|-Value	Probability $y > \|t\|$
A	3.647197275	3.647197275	0.002383498
B	-0.453542398	0.453542398	0.656650632
C	-1.939143963	1.939143963	0.071531183
D	0.92249295	0.92249295	0.370872546
E	-2.012759003	2.012759003	0.062450281
F	-0.273389668	0.273389668	0.788282063
G	-3.147931904	3.147931904	0.006634361
H	1.289935684	1.289935684	0.21660624
I	3.722260912	3.722260912	0.002043765
J	1.66525387	1.66525387	0.11660421
K	-0.050569187	0.050569187	0.960336071
L	-1.35325252	1.35325252	0.196018828
M	0.701278739	0.701278739	0.493866038
N	0.624871858	0.624871858	0.541444728
O	2.614716698	2.614716698	0.019518401

TABLE 4.12

Values of Percentage Confidence Level

Variables	% Confidence Level	Variables	% Confidence Level	Variables	% Confidence Level
A	99.76165023	F	21.17179373	K	3.966392914
B	34.33493683	G	99.33656393	L	80.39811722
C	92.84688171	H	78.33937596	M	50.61339622
D	62.91274536	I	99.79562347	N	45.85552719
E	93.75497187	J	88.33957904	O	98.0481599

Calculation of confidence level

$$\text{\% Confidence level} = (1 - \text{Probability due to chance}) \times 100$$

For example, % confidence level for variable E = $(1–0.062450281) \times 100$ = 93.75497187

Other values of % confidence level are in Table 4.12.

Based upon the confidence level Table 4.12, the average confidence level is 69.9650477%.

Therefore, the confidence level cut-off value may be considered as 10% above the average confidence level value, that is, at least above 80%.

Based on the confidence level cut-off value, variables A, C, E, G, I, J, L, and O are screened for further experiments. However, O is a dummy variable.

In order to validate the screened variables, the sum of squares value for each variable has been calculated.

Calculation of sum of squares
For variable E

$$\text{Sum of square} = \left[\frac{(\text{Sum of positive levels})^2}{8}\right] + \left[\frac{(\text{Sum of negative levels})^2}{8}\right] - \left[\frac{(\text{Total sum of levels})^2}{16}\right]$$

$$= (1.367445 + 1.631902) - 2.993506$$

$$= 0.005841$$

Table 4.13 gives the values of sum of squares of other variables.

Sum of squares values for the selected variables exactly match with the percentage confidence levels.

For the first-order design, the first-order model (Equation 3.1) is considered here. There is no interaction term. Linear effect terms and offset term are calculated in this example (Table 4.14). Calculation will be discussed in the Appendix A.2.

TABLE 4.13

Values of Sum of Squares of Variables

| Variables | Sum of Squares | | | | |
	S(+ level)²/8	S(-level)²/8	Total	(S(Y))²/16	Sum of Squares
A	−1.745926	1.266754	3.012681	2.993506	0.019175
B	−1.467099	1.526703	2.993802	2.993506	0.000297
C	−1.691696	1.626847	3.318543	2.993506	0.325038
D	−1.557966	1.436767	2.994732	2.993506	0.001227
E	1.367445	1.631902	2.999346	2.993506	0.005841
F	1.478856	1.514757	2.993613	2.993506	0.000108
G	1.297097	1.710695	3.007792	2.993506	0.014286
H	1.582687	1.413217	2.995904	2.993506	0.002399
I	1.751256	1.262222	3.013478	2.993506	0.019973
J	1.608142	1.389361	2.997503	2.993506	0.003997
K	1.493424	1.500085	2.993509	2.993506	0.000004
L	1.409185	1.586960	2.996145	2.993506	0.002639
M	1.543173	1.451041	2.990214	2.993506	0.000709
N	1.538083	1.455986	2.944068	2.993506	0.000563
O	1.67344	1.329917	3.00361	2.993506	0.009856

TABLE 4.14

Values of Linear Coefficients

Variables	Coefficient (β)
Mean	$\beta_0 = 0.4325$
A	0.0346
B	−0.0043
C	−0.0184
D	0.0088
E	−0.0191
F	−0.0026
G	−0.0229
H	0.0122
I	0.0353
J	0.0158
K	−0.0005
L	−0.0128
M	0.0067
N	0.0059
O	0.0248

Example 4.3

Screening of medium components for endoglucanase production by a fungus in shake culture was desired. Fermentation conditions were volume of the enzyme production medium – 100 cm^3 in 500 cm^3 Erlenmeyer flask. Constituents of the unoptimized medium are in kg/m^3 major carbon source 10; other medium constituents were added as given in the Table 4.15. Age of the organism in slant growth – 120 h, inoculums age – 36 h, inoculums level – 5% (v/v) (0.24 g/l dry equivalent of cells approximately), shaker speed – 180 rpm, fermentation time – 216 h temperature of fermentation 30°C, initial pH of the enzyme production medium – 5 and pH uncontrolled during fermentation. Experimental variables of this fermentation are in Table 4.15.

Analyze the statistical parameters for the proposed Plackett and Burman design. Table 4.16 is the design matrix.

Solution:

The procedure discussed above for the Plackett and Burman design is applicable in this case. Variables as medium constituents, their effects, coefficients, sum of squares, *t*-values, probability due to chance, and confidence levels obtained by analyzing the experimental data for the Plackett and Burman design used for screening of medium components for enzyme production are shown in Table 4.17.

Decision

Urea and $ZnSO_4 \cdot 7H_2O$ have very low percentage confidence level. They are not considered for optimization studies.

TABLE 4.15

Variables and Their Range

Code for Variables	Variables	(+) Level at (kg/m³)	(–) Level at (kg/m³)
A	KH_2PO_4	2.5	2
B	$(NH_4)_2 SO_4$	1.75	1.4
C	$NaH_2PO_4 \cdot 2H_2O$	9.75	7.8
D	$MgSO_4 \cdot 7H_2O$	0.037	0.03
E	Peptone	12	9.6
F	Citric acid (anhydrous)	0.25	0.2
G	Tween 80	0.375	0.2
H	Urea	1.25	0.1
I	$FeSO_4 \cdot 7H_2O$	0.00625	0.005
J	$ZnSO_4 \cdot 7H_2O$	0.00129	0.001
K	$MnSO_4 \cdot H_2O$	0.00175	0.0014
L	$CaCl_2 \cdot 2H_2O$	0.0025	0.002
M, N, and O	Dummy variables	–	–

TABLE 4.16

Design Matrix with the Response

Run No.	A	B	C	D	E	F	G	H	I	J	K	L	M	N	O	Enzyme (U) as the Response
1	+	+	+	+	−	+	−	+	+	−	−	+	−	−	−	1.43
2	−	+	+	+	+	−	+	−	+	+	−	−	+	−	−	1.29
3	−	−	+	+	+	+	−	+	−	+	+	−	−	+	−	1.30
4	−	−	−	+	+	+	+	−	+	−	+	+	−	−	+	0.91
5	+	−	−	−	+	+	+	+	−	+	−	+	+	−	−	1.01
6	−	+	−	−	−	+	+	+	+	−	+	−	+	+	−	0.84
7	−	−	+	−	−	−	+	+	+	+	−	+	−	+	+	0.93
8	+	−	−	+	−	−	−	+	+	+	+	−	+	−	+	0.99
9	+	+	−	−	+	−	−	−	+	+	+	+	−	+	−	1.29
10	−	+	+	−	−	+	−	−	−	+	+	+	+	−	+	1.26
11	+	−	+	+	−	−	+	−	−	−	+	+	+	+	−	1.31
12	−	+	−	+	+	−	−	+	−	−	−	+	+	+	+	1.13
13	+	−	+	−	+	+	−	−	+	−	−	−	+	+	+	1.25
14	+	+	−	+	−	+	+	−	−	+	−	−	−	+	+	0.98
15	+	+	+	−	+	−	+	+	−	−	+	−	−	−	+	1.50
16	−	−	−	−	−	−	−	−	−	−	−	−	−	−	−	0.69

TABLE 4.17

Statistical Parameters for the Design Matrix in Table 4.16

Variables	Effects	Coefficient β_i	Sum of Squares	t for H_0 Parameter = 0	Probability > $|t|$	Confidence Level, %
Mean	−	−	−	36.116	0.0001	−
A	0.1756	0.0878	0.1233	10.549	0.0018	99.82
B	0.1666	0.0833	0.1110	10.010	0.0021	99.79
C	0.3024	0.1512	0.3657	18.166	0.0004	99.96
D	0.0700	0.0350	0.0196	4.205	0.0246	97.54
E	0.1571	0.0785	0.0987	9.438	0.0025	99.75
F	−0.0191	−0.0095	0.0015	−1.147	0.3346	66.54
G	−0.0715	−0.0357	0.0204	−4.293	0.0232	97.68
H	0.0175	0.0087	0.0012	1.051	0.3706	62.94
I	−0.0333	−0.0167	0.0044	−2.001	0.1391	86.09
J	−0.0019	−0.0010	−	−0.116	0.9147	8.53
K	0.0885	0.0443	0.0313	4.318	0.0130	98.70
L	0.0533	0.0266	0.0113	3.200	0.0493	95.07

4.2.2 Second-Order Designs

4.2.2.1 Central-Composite Designs

The concept of the different CCDs and the design matrices are described in the Chapter 3. Analysis of this design begins with the construction of the ANOVA table, followed by the calculation of optimal conditions of the experimental variables.

Construction of the ANOVA table
Experiments have been carried out according to the central-composite-experimental-design plan, described in Chapter 3. Analyses of data, followed by the calculation of statistical components, are given in a tabular form. The table is called *ANOVA table*. The entries in the table give the information regarding the separate sources of variation in the data.

The total variation in a set of data is called the *total sum of squares* (SST). The quantity SST is the sum of the squares of the deviations of the observed response (Y_0) about their average value for N observations.

$$\overline{Y} = \frac{(Y_1 + Y_2 + Y_3 + \cdots + Y_N)}{N}$$

$$\text{SST} = \sum_{o=1}^{N} (Y_o - \overline{Y})^2 \tag{4.16}$$

The quantity SST has associated with $(N–1)$ degrees of freedom since the sum of deviations $(Y_o - \overline{Y})$ is equal to zero. The SST is of two parts; the sum of squares due to regression (SSR) (or the sum of squares explained by the fitted model) and the sum of squares unaccounted for by the fitted model. The sum of squares due to regression SSR is given in Equation 4.17.

$$\text{SSR} = \sum_{o=1}^{N} (\tilde{Y}(x_o) - \overline{Y})^2 \tag{4.17}$$

The deviation $(\tilde{Y}(x_o) - \overline{Y})^2$ is the difference between the value predicted by the fitted model for the 0th observation and the overall average of the Y_o's. If the fitted model, where $\tilde{Y}$ is the predicted response, β_0 is the offset term, β_i is the linear effect term, β_{ii} is the squared effect term, β_{ij} is the interaction terms, and x_i x_j are the coded values of independent variables, contains p parameters. SSR is associated with $(p - 1)$ degrees of freedom. SSE is the sum of squares unaccounted for by the fitted model or sum of squares of residuals or sum of squares of the errors, is Equation 4.18, having degrees of freedom $(N - 1) - (p - 1) = (N - p)$.

$$\text{SSE} = \sum_{o=1}^{N} (Y_o - \tilde{Y}(x_o))^2 \tag{4.18}$$

The usual test of significance of the fitted regression equation is a test of null hypothesis (H_0). The null hypothesis is: all values of β_i (excluding β_0) is zero. If normal distribution of errors is assumed, H_0 test involves calculation of F-statistic (Equation 4.19).

$$F = \frac{\text{SSR}/(p-1)}{\text{SSE}/(N-p)} = \frac{\text{Mean square regression}}{\text{Mean square residual}} \tag{4.19}$$

What shall one do with the F-value?
   If H_0 is true,

1. F-statistic follows F-distribution with $(p-1)$ df in the numerator and $(N-p)$ df in the denominator.
2. To compare the value of F-statistic from Equation 4.19 to the value of $F_{\alpha,(p-1),(N-p)}$, which is the upper 100α percent point of the F-distribution with $(p-1)$ and $(N-p)$ df, being obtained from F-distribution chart. F-distribution chart is available from standard book on statistics.

Decision from the analysis of F-statistic
This is based on the acceptance or rejection of the hypothesis, H_0 or $H_{\text{alternative}}$.
   The alternative hypothesis is H_a: atleast one value of β_i (excluding β_0) is not zero.

Example 4.4

Formulate the ANOVA table and calculate F-statistic and R^2 from the following data in Table 4.18.

TABLE 4.18

Variables and Responses

Run Number	Variable 1 (x_1)	Variable 2 (x_2)	Observed Response (Y)	Deviation $(Y-\bar{Y})$	Predicted Response $(\tilde{Y})$	Deviation $(\tilde{Y}-\bar{Y})$	Residual $(Y-\tilde{Y})$
1	−1	−1	0.3367	−0.0127	0.3320	−0.0174	0.0047
2	1	−1	0.3448	−0.0046	0.3440	−0.0054	0.0008
3	−1	1	0.2587	−0.0907	0.2607	−0.0887	−0.0020
4	1	1	0.2728	−0.0766	0.2788	−0.0706	−0.0060
5	−1	0	0.3536	0.0042	0.3563	0.0069	−0.0027
6	1	0	0.3766	0.0272	0.3714	0.0220	0.0052
7	0	−1	0.3571	0.0077	0.3626	0.0132	−0.0055
8	0	1	0.3023	−0.0471	0.2943	−0.0551	0.0080
9	0	0	0.3914	0.0420	0.3884	0.0390	0.0030
10	0	0	0.3817	0.0323	0.3884	0.0390	−0.0067
11	0	0	0.3892	0.0398	0.3884	0.0390	0.0008
12	0	0	0.3921	0.0427	0.3884	0.0390	0.0037
13	0	0	0.3852	0.0358	0.3884	0.0390	−0.0032

TABLE 4.19

Components of ANOVA

Sources of Variation	Degrees of Freedom (df)	Sum Squares (SS)	Mean Squares (MS)	F-Value	Probability > F
Due to fitted model(regression)	$(p-1)$	SSR	$SSR/(p-1)$	$\dfrac{MSR}{MSE}$	From standard F-distribution table
Residual (error)	$(N-p)$	SSE	$SSE/(N-p)$		
Total	$N-1$	SST			

Solution:

Construction of ANOVA: The template of ANOVA is Table 4.19.

Hence, from Table 4.18, calculation of entries in Table 4.19 is given below.

$$\text{Number of observation} = N = 13$$

$$\bar{Y} = \frac{4.5422}{13} = 0.3494$$

$$SST = \sum_{o=1}^{13}(Y_o - 0.3494)^2 = 0.024808 \text{ with } (13-1) = 12df$$

$$SSR = \sum_{o=1}^{13}(\tilde{Y}(x_o) - 0.3494)^2 = 0.02453 \text{ with } (6-1) = 5df$$

$$SSE = \sum_{o=1}^{13}(Y_o - \tilde{Y}(x_o))^2 = SST - SSR = 0.0002695 \text{ with } (13-6) = 7df$$

$$\text{Mean square regression} = \frac{0.02453}{5} = 0.004906$$

$$\text{Mean square residual} = \frac{0.0002695}{7} = 0.0000385$$

$$F = \frac{MSR}{MSE} = \frac{0.004906}{0.0000385} \approx 127$$

$F_{tabulated}$ at a probability level of 0.0001 is < 127.

For this example, ANOVA table is Table 4.20.

Decision:

H_0 is rejected at the 0.0001 level of significance. For $H_{alternative}$ at least one of the parameters, other than β_0, in the fitted model is not zero.

TABLE 4.20

ANOVA

Sources of Variation	Degrees of Freedom (df)	Sum Squares (SS)	Mean Squares (MS)	F-Value	Probability > F Probability, 0.0001
Due to fitted model(regression)	$(p-1) = 5$	SSR = 0.02453	SSR/(p – 1) 0.004906	MSR/ MSE	From standard F-distribution
Residual (error)	$(N - p) = 7$	SSE = 0.0002695	SSE/(N – p) = 0.0000385	≈ 127	table
Total	$N - 1 = 12$	SST = 0.024808			

Example 4.5

Effect of medium constituents on the production of extracellular enzyme has been studied to find the optimal conditions of carbon source, ammonium nitrate, malt extract and urea. CCD experimental plan is suggested in this study. Table 4.21 gives the design plan and the enzyme activity as the response. Justify the class of CCD. Calculate the model parameters, if second-order regression model represent the response. Calculate the entries in the ANOVA table and R^2.

Solution:

> Total number of runs = 30 = N
> Number of factors = 4 = k
> Total number of level = 2
> Number of axial experimental points = $2k$ = 8 points
> Total number of design points = $N = 2^k + 2k + n_c \cdot n_c$ = Number of centre points in the design. $n_c = 30 - 16 - 8 = 6$

There are six centre points. This can be verified by counting the centre points from the design matrix. For simplicity, the variables in coded form are given below.

$$x_i = \frac{2(X_i - \overline{X_i})}{R_i}$$

For x_1, $\overline{X_i}$ = Mean of X_is = 5, R_i = Range of X_is = 5

$$\alpha = 2\frac{(10 - 5)}{5} = 2$$

For CCD to be orthogonal,

$$\alpha = \left[\frac{(FN)^{1/2} - F}{2} \right]^{1/2} = \left[\frac{(16 \times 30)^{1/2} - 16}{2} \right]^{1/2} = 1.72$$

TABLE 4.21

Variables with the Response for Enzyme Synthesis

Run No.	C-Source (kg/m³) X_1	Ammonium Nitrate (kg/m³) X_2	Malt Extract (kg/m³) X_3	Urea (kg/m³) X_4	Enzyme (U) Experimental (Y_3) Set 1	Set 2	Enzyme (U) Predicted (Y_3)
1	2.5	0.7	0.5	0.45	0.0702	0.0650	0.0572
2	7.5	0.7	0.5	0.15	0.1000	0.1156	0.1018
3	2.5	2.1	0.5	0.15	0.0732	0.1256	0.0908
4	7.5	2.1	0.5	0.45	0.0559	0.0929	0.0729
5	2.5	0.7	1.5	0.15	0.0570	0.0554	0.0545
6	7.5	0.7	1.5	0.45	0.0572	0.0580	0.0631
7	2.5	2.1	1.5	0.45	0.0376	0.0420	0.0427
8	7.5	2.1	1.5	0.15	0.0558	0.0546	0.0626
9	5.0	1.4	1.0	0.3	0.0566	0.0566	0.0636
10	5.0	1.4	1.0	0.3	0.0566	0.0606	0.0636
11	2.5	0.7	0.5	0.15	0.0814	0.0950	0.0784
12	7.5	0.7	0.5	0.45	0.0764	0.0844	0.0725
13	2.5	2.1	0.5	0.45	0.0612	0.0616	0.0561
14	7.5	2.1	0.5	0.15	0.1044	0.1112	0.1014
15	2.5	0.7	1.5	0.15	0.0430	0.0430	0.0461
16	7.5	0.7	1.5	0.15	0.0616	0.0732	0.0673
17	2.5	2.1	1.5	0.15	0.0520	0.0392	0.0504
18	7.5	2.1	1.5	0.45	0.0381	0.0428	0.0469
19	5.0	1.4	1.0	0.3	0.0468	0.0476	0.0606
20	5.0	1.4	1.0	0.3	0.0610	0.0610	0.0606
21	0.0	1.4	1.0	0.3	0.0118	0.0130	0.0206
22	10	1.4	1.0	0.3	0.0530	0.0474	0.0483
23	5.0	0	1.0	0.3	0.0404	0.0308	0.0461
24	5.0	2.8	1.0	0.3	0.0502	0.0430	0.0423
25	5.0	1.4	0	0.3	0.0532	0.0444	0.0725
26	5.0	1.4	2.0	0.3	0.0402	0.0402	0.0277
27	5.0	1.4	1.0	0	0.0448	0.0688	0.0637
28	5.0	1.4	1.0	0.6	0.0548	0.0294	0.0267
29	5.0	1.4	1.0	0.3	0.0404	0.0404	0.0383
30	5.0	1.4	1.0	0.3	0.0608	0.0620	0.0383

For CCD to be rotatable,

$$\alpha = F^{1/4} \Rightarrow \alpha_{rot} = 16^{1/4} = 2$$

For CCD to be spherical,

$$\alpha = \sqrt{k} \Rightarrow \alpha_{sph} = \sqrt{4} = 2$$

Thus, the given orthogonal design is a rotatable as well as spherical.
Estimated number of centre points,

$$\alpha = \alpha_{rotatable} = \alpha_{spherical} = 2, \alpha_{orthogonal} \neq \alpha_{actual}$$

Estimated number of centre points

$$n_0 \approx 4\sqrt{F} + 4 - 2k$$

$$n_0 \approx 12 \text{ centre points}$$

Only six centre points have been used, which is lesser than what is expected by calculation (i.e. half the number). So, the given design matrix has insufficient number of centre points.

The regression model for the response is Equation 4.20.

$$\tilde{Y} = \beta_0 + \beta_1 x_1 + \beta_2 x_2 + \beta_3 x_3 + \beta_4 x_4 + \beta_{12} x_1 x_2 + \beta_{13} x_1 x_3$$
$$+ \beta_{14} x_1 x_4 + \beta_{23} x_2 x_3 + \beta_{24} x_2 x_4 + \beta_{34} x_3 x_4 \qquad (4.20)$$
$$+ \beta_{11} x^2_1 + \beta_{22} \beta_1 x_1 + \beta_{33} x^2_3 + \beta_{44} x^2_4$$

The solution to the above set of β values is got by solving the matrix system,

$$\beta = \left(X^T X\right)^{-1} \left(X^T Y\right) \qquad (4.21)$$

where:
Y is the matrix with response values

We get β_i.

$\beta_0 = 0.0542$	$\beta_{12} = -0.0022$	$\beta_{34} = 0.0043$
$\beta_1 = 0.0073$	$\beta_{13} = -0.0006$	$\beta_{11} = -0.0011$
$\beta_2 = -0.0006$	$\beta_{14} = 0$ (negligible)	$\beta_{22} = 0.0014$
$\beta_3 = -0.0128$	$\beta_{23} = -0.0021$	$\beta_{33} = 0.0022$
$\beta_4 = -0.0095$	$\beta_{24} = -0.0009$	$\beta_{44} = 0.0016$

For our given dataset

$$SST = \sum_{o=1}^{N} \left(Y_o - \overline{Y}\right)^2 = 0.0141, df = 29$$

$$SSR = \sum_{o=1}^{N} \left(Y_{(x_o)} - \overline{Y}\right)^2 = 0.0078, df = 15 - 1 = 14$$

$$SSE = 0.0141 - 0.0078 = 0.0063, df = N - p = 30 - 15 = 15$$

$$R^2 = \frac{SSR}{SST} = 0.5532$$

So, the given system has an R^2 value of 0.5532.

Example 4.6

CCD matrix and corresponding enzyme production by a fungus having conditions: volume of the enzyme production medium – 100 cm³ in 500 cm³ Erlemeyer flask, age of the organism – 120 h, inoculums age – 36 h, inoculum level –0.24 kg/m³ dry equivalent of cells approximately, fermentation time – 216 h, shaker speed – 180 rpm, temperature of fermentation – 30°C, initial pH of the enzyme production medium – 5. The values for the concentrations of phosphate, ammonium nitrate, malt extract, surfactant and minerals, which are variables in this experiment, are given in Tables 4.22 and 4.23. The rest of the medium compounds are dried grass 10; citric acid anhydrous, 9.6; urea, 0.3; $ZnSO_4 \cdot H_2O$, 0.0014. Maximum enzyme production was achieved at 192 h of fermentation.

TABLE 4.22

Variables with the Response for Enzyme Synthesis

Run No.	Phosphate x_1	Ammonium Nitrate, x_2	Malt Extract x_3	Surfactant x_4	Minerals x_5	Enzyme, U Experimental	Enzyme, U Predicted
1	−1	−1	−1	−1	1	0.56	0.7058
2	1	−1	−1	−1	−1	1.77	1.7702
3	−1	1	−1	−1	−1	0.22	0.3198
4	1	1	−1	−1	1	1.69	1.8741
5	−1	−1	1	−1	−1	0.02	−0.0818
6	1	−1	1	−1	1	0.75	0.7338
7	−1	1	1	−1	1	0.01	0.0956
8	1	1	1	−1	−1	0.82	0.7581
9	−1	−1	−1	1	−1	0.22	0.1372
10	1	−1	−1	1	1	1.37	1.3730
11	−1	1	−1	1	1	0.04	0.1428
12	1	1	−1	1	−1	0.25	0.2094
13	−1	−1	1	1	1	0.25	0.1430
14	1	−1	1	1	−1	1.12	0.8662
15	−1	1	1	1	−1	0.02	−0.1237
16	1	1	1	1	1	0.37	0.3066
17	−2	0	0	0	0	0	−0.0350
18	2	0	0	0	0	1.51	1.6032
19	0	−2	0	0	0	0.32	0.4865
20	0	2	0	0	0	0.08	−0.0297
21	0	0	−2	0	0	1.52	1.2807
22	0	0	2	0	0	0.02	0.3221
23	0	0	0	−2	0	1.32	0.8720
24	0	0	0	2	0	0.08	1.2519
25	0	0	0	0	−2	0.61	1.2975
26	0	0	0	0	2	1.45	1.2975

(Continued)

TABLE 4.22 (*Continued*)

Variables with the Response for Enzyme Synthesis

Run No.	Phosphate x_1	Ammonium Nitrate, x_2	Malt Extract x_3	Surfactant x_4	Minerals x_5	Enzyme, U Experimental	Enzyme, U Predicted
27	0	0	0	0	0	1.21	1.2975
28	0	0	0	0	0	1.28	1.2975
29	0	0	0	0	0	1.45	1.2975
30	0	0	0	0	0	1.14	1.2975
31	0	0	0	0	0	1.39	1.2975
32	0	0	0	0	0	1.37	1.2975

TABLE 4.23

Variables with Their Levels

Levels	X_1	X_2	X_3	X_4	X_5
$-\alpha$	0	0	0	0	0
-1	3.84	0.88	0.63	0.125	0.0196
0	7.68	1.75	1.25	0.25	0.039
$+1$	11.52	2.63	1.88	0.375	0.059
$+\alpha$	15.36	3.50	2.5	0.50	0.079

TABLE 4.24

ANOVA

Variation due to	Sum Squares	Degrees of Freedom	Mean Square	F-Value	Probability > F
Model	11.23359	20	0.561680	8.874	0.0003
Error	0.69621	11	0.063292		
Corrected total	11.92980	31	–		

Solution:

Coding of variables: Table 4.23 represents coded and decoded values in kg/m³ of variables.

ANOVA obtained from CCD employed in optimization of the enzyme production is given in Table 4.24.

Entries in ANOVA table are calculated as per the procedure and formulae described earlier.

The second-order polynomial equation thus obtained for enzyme production is as follows:

$$\tilde{Y} = 1.2975 + 0.4095x_1 - 0.1291 x_2 - 0.2397 x_3 - 0.1951 x_4$$
$$+ 0.095 x_5 - 0.1284 x_1^2 - 0.2673 x_2^2 - 0.124 x_3^2$$
$$- 0.1421 x_4^2 - 0.0589 x_5^2 - 0.0703 x_1x_2 - 0.0806x_1x_3 \quad (4.22)$$
$$- 0.1026 x_1x_4 - 0.0095 x_1x_5 + 0.051 x_2x_3 - 0.119 x_2x_4$$
$$+ 0.062 x_2x_5 + 0.1559 x_3x_4 - 0.1124 x_3x_5 + 0.0146 x_4x_5$$

where:

$\tilde{Y}$ is the predicted response

x_i is the coded value of the concentration of variables as given in Table 4.23

The coefficients of this equation are calculated using the method described earlier (refer Appendix A.2).

4.2.2.2 Box-Behenken Design

The example is incomplete block design for three-level experiments. This is achieved by combining two-level factorial design with the block effect.

For m factors, the number of block points $= 4 \times {}^m c_2 = 2m(m-1)$. We have three factors. Number of block points $= 3 \times 2 \times 2 = 12$. Experimental problem has 15 points out of which three points are the centre points. The remaining points constitute the Box-Behenken complete block design.

1. The pH has three levels 6, 6.5, and 7, which are transformed to equation.

$$x_1 (pH) = \frac{pH \text{ value} - 6.5}{0.5}$$

as –1, 0, and +1, respectively.

2. Aeration rate (m³ air/m³ of reaction fluid [min]) has three levels 1, 1.5, and 2, which are transferred to

$$x_2 (\text{Aeration rate}) = \frac{\text{Aeration rate} - 1.5}{0.5}$$

–1, 0, and 1, respectively.

3. Agitation rate (rev/min) has three levels 100, 160, and 220, which are transferred as

$$x_3 (\text{Agitation rate}) = \frac{\text{Agitation rate} - 160}{60}$$

–1, 0, and 1, respectively.

Now, in the design matrix three factors have three levels in Box and Behenken design, having three centre points. In Table 4.25, the design order gives the composition of the matrix for Box and Behenken design.

TABLE 4.25

Experiments as per the Box and Behnken Design with the Response Produced by *Aspergillus niger* in a Batch Bioreactor

Run Number	Design Points	pH	Aeration (m³/m³ (min))	Agitation (rev/min)	Fermentation Time (h)	Yield of		
						Organic Acid (kg/kg Reactant)	Lactone Form of Organic Acid (kg/kg Reactant)	Mycelia (kg dry wt./kg Reactant)
1	1	−1(≡6)	−1(≡1)	0(≡160)	24	0.205	0.04	0.083
					48	0.405	0.05	0.210
2	3	+1(≡7)	−1(≡1)	0(≡160)	24	0.26	0.09	0.070
					48	0.28	0.02	0.239
3	5	0(≡6.5)	0(≡1.5)	0(≡160)	24	0.19	0.06	0.053
					48	0.38	0.05	0.112
4	8	+1(≡7)	0(≡1.5)	−1(≡100)	24	0.315	0.08	0.090
					48	0.36	0.066	0.242
5	14	0(≡6.5)	+1(≡2)	+1(≡220)	24	0.17	0.01	0.116
					48	0.438	0.015	0.150
6	6	−1(≡6)	0(≡1.5)	−1(≡100)	24	0.24	0.021	0.102
					48	0.37	0.030	0.143
7	7	−1(≡6)	0(≡1.5)	+1(≡220)	24	0.257	0.10	0.090
					48	0.36	0.067	0.110
8	13	0(≡6.5)	+1(≡2)	−1(≡100)	24	0.26	0.028	0.070
					48	0.35	0.015	0.138
9	9	+1(≡7)	0(≡1.5)	+1(≡220)	24	0.20	0.11	0.107
					48	0.37	0.09	0.125
10	12	0(≡6.5)	−1(≡1)	+1(≡220)	24	0.23	0.04	0.083
					48	0.36	0.021	0.120

(Continued)

TABLE 4.25 (*Continued*)

Experiments as per the Box and Behnken Design with the Response Produced by *Aspergillus niger* in a Batch Bioreactor

Run Number	Design Points	pH	Aeration (m^3/m^3 (min))	Agitation (rev/min)	Fermentation Time (h)	Yield of Organic Acid (kg/kg Reactant)	Lactone Form of Organic Acid (kg/kg Reactant)	Mycelia (kg dry wt./kg Reactant)
11	4	+1(≡7)	+1(≡2)	0(≡160)	24	0.30	0.06	0.100
					48	0.417	0.058	0.141
12	10	0(≡6.5)	0(≡1.5)	0(≡160)	24	0.182	0.056	0.093
					48	0.385	0.51	0.160
13	11	0(≡6.5)	−1(≡1)	−1(≡100)	24	0.19	0.09	0.060
					48	0.38	0.08	0.117
14	15	0(≡6.5)	0(≡1.5)	0(≡160)	24	Nil	Nil	Nil
					48	Nil	Nil	Nil
15	2	−1(≡6)	+1(≡2)	0(≡160)	24	0.15	0.09	0.081
					48	0.34	0.07	0.133

Note: The response for organic acid produced at 48 h is only used for calculation of statistical parameters

For, yield at 48 h, a quadratic model fit is the Equation (4.23).

$$Y = \beta_0 + \beta_1 x_1 + \beta_2 x_2 + \beta_3 x_3 + \beta_{11} x_1^2 + \beta_{22} x_2^2$$
$$+ \beta_{33} x_3^2 + \beta_{12} x_1 x_2 + \beta_{23} x_2 x_3 + \beta_{31} x_3 x_1 \tag{4.23}$$

This is a 10-parameter model, which is represented in the following form.

$$Y = X\beta \tag{4.24}$$

Solve for β using linear least squares as

$$\beta = \left(X^T X\right)^{-1} X^T Y \tag{4.25}$$

Proper substitution of the values,

$$\beta = \begin{vmatrix} \beta_0 \\ \beta_1 \\ \beta_2 \\ \beta_3 \\ \beta_{11} \\ \beta_{22} \\ \beta_{33} \\ \beta_{12} \\ \beta_{23} \\ \beta_{31} \end{vmatrix} = \begin{vmatrix} 0.3825 \\ -0.006 \\ 0.015 \\ 0.0085 \\ -0.0195 \\ -0.0025 \\ 0.0020 \\ 0.0505 \\ 0.0270 \\ 0.005 \end{vmatrix}$$

More effects from $|\beta|$ are β_{12}, β_{23}, and β_2, that is, the interaction effects of $(x_1$ and $x_2)$, $(x_2$ and $x_3)$ and the linear effect of x_2, respectively.

Predicted values
All required calculations are given in Table 4.26.
   Equation 4.26 is used to get the response surface.

$$\hat{Y} = \hat{\beta} X \tag{4.26}$$

$R^2 = SS_{explained}/SS_{total} = 0.9477$, which is the same as the reported values.

4.3 To Find Optimal Conditions of Experimental Variables for the Bioprocesses

The discussion is limited to the single and multiple responses. For single response, the response may be a linear function or curvilinear functions. Response from OFAT design without interaction terms in the model equation

TABLE 4.26

Calculation for Predicted Values of Response

β	x_1	x_2	x_3	x_1^2	x_2^2	x_3^2	x_1x_2	x_2x_3	x_3x_1	$Y_{experimental}$	$Y_{predicted}$	β	
1	−1	−1	0	1	1	0	1	0	0	0.405	0.402	0.3825	$=\beta_0$
1	1	−1	0	1	1	0	−1	0	0	0.28	0.289	−0.006	$=\beta_1$
1	1	1	0	1	1	0	1	0	0	0.417	0.42	0.015	$=\beta_2$
1	−1	1	0	1	1	0	−1	0	0	0.34	0.331	0.0085	$=\beta_3$
1	1	0	−1	1	0	1	0	0	−1	0.36	0.345	−0.0195	$=\beta_{11}$
1	−1	0	−1	1	0	1	0	0	1	0.37	0.367	−0.0025	$=\beta_{22}$
1	−1	0	1	1	0	1	0	0	−1	0.36	0.374	0.002	$=\beta_{33}$
1	1	0	1	1	0	1	0	0	0	0.37	0.372	0.0505	$=\beta_{12}$
1	0	1	−1	0	1	1	0	−1	0	0.35	0.361	0.027	$=\beta_{23}$
1	0	−1	−1	0	1	1	0	1	0	0.38	0.385	0.005	$=\beta_{31}$
1	0	−1	1	0	1	1	0	−1	0	0.36	0.348		
1	0	1	1	0	1	1	0	1	0	0.438	0.432		
1	0	0	0	0	0	0	0	0	0	0.385	0.382		
1	0	0	0	0	0	0	0	0	0	0.38	0.382		
1	0	0	0	0	0	0	0	0	0	0	0.382		

will be linearly correlated to the experimental variables. In other cases, response variable will be correlated to the experimental variables involving linear, interaction, and quadratic terms. In multi-response analysis, the discussions will curvilinear nature.

4.3.1 For a First-Order Surface

The path of steepest ascent[4]

Method: The steps and the technique are described in detail by Khuri and Conell.[4]

Step 1: One needs to perform a number of logical experiments in the direction of increased response. The direction depends on the scale of coded variables.

Step 2: An appropriate equation of hyperplane gives the idea of approximate response surface. From the information of hyperplane, a direction of increased response is assumed.

Step 3: In this increased direction, the experimenter will continue the experiments until a curvature. If an experiment continued in this direction causes a decreased performance, again a first-order model is fitted. If in a new direction increased response appears, further logical experiments continue until there is no increase in response. Figure 4.3 gives concept of the method.

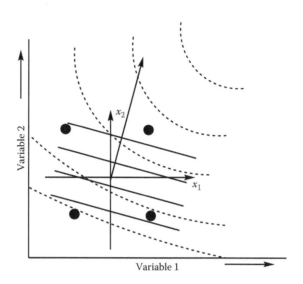

FIGURE 4.3
Concept of steepest ascent.

4.3.2 For a Second-Order Surface

Separate discussion is for single response and multi-response systems.

4.3.2.1 For Single-Response System

The two techniques are derivative and derivative-free, which are discussed here.

4.3.2.1.1 Derivative Technique

Determination of optimal values of the variables by solving the regression model.

Optimization of the regression equation obtained after the statistical analysis of the experimental data is done by a program written in FORTRAN using the algorithms.[6–8]

Here, the algorithm of Rosenbrock (see Sections A.5 and A.6 in Appendix) is used to maximize the function.

$$\hat{Y}_i = \beta_0 + \sum \beta_i X_i + \sum \beta_i X_i^2 + \sum \beta_{ij} X_i X_j \tag{4.27}$$

subject to the constraints $-\alpha \le x_i \le +\alpha$; $i = 1, 2, 3$, where α is the distance of the axial point from the design centre in the central-composite experimental design. However, when the optimum value of an independent variable lies either on the lower boundary or on the higher boundary of the zone under investigation, the regression equation is optimized taking the constraint limits beyond the selected values. If the optimum value of an independent variable (obtained by optimizing the regression equation) does not remain in the experimental domain or if it falls on the boundary, a separate experiment taking into account the optimal values of the independent variables is conducted to justify the optima. The program listed for the algorithm of Rosenbrock is given in Appendix A.5 for the response function Equation 4.36. A logic diagram showing the Rosenbrock procedure to handle constraints is given in Appendix A.5.

$$\hat{Y} = 0.03471 + 0.002703x_1 - 0.017152x_2 - 0.010682x_1^2$$
$$+ 0.01122x_2^2 - 0.001788x_1x_2$$

(4.28)

Example 4.6

Equation 4.29 is the regression equation developed for medium composition optimization.

$$\hat{Y} = \left(1.67 \times 10^{-3}\right) + \left(1.84 \times 10^{-4}x_1\right) + \left(9.257 \times 10^{-5}x_2\right)$$
$$+ \left(1.566 \times 10^{-5} x_3\right) + \left(1.504 \times 10^{-5} x_4\right)$$
$$- \left(1.798 \times 10^{-4}x_1^2\right) - \left(9.431 \times 10^{-5} x_2^2\right)$$
$$- \left(1.087 \times 10^{-4} x_3^2\right) - \left(5.83 \times 10^{-5} x_4^2\right)$$
$$+ \left(9.40 \times 10^{-5} x_1x_2\right) - \left(9.489 \times 10^{-5} x_1x_3\right)$$
$$+ \left(1.723 \times 10^{-5} x_1x_4\right) + \left(4.633 \times 10^{-5} x_2x_3\right)$$
$$- \left(2.72 \times 10^{-5} x_2x_4\right) + \left(2.426 \times 10^{-5} x_3x_4\right)$$

(4.29)

where:
$\hat{Y}$ is the predicted response. $x_1, x_2, x_3,$ and x_4 are controlled variables

To solve Equation 4.29, it is first partially differentiated with respect to x_1, $x_2, x_3,$ and x_4 which is followed by equating it to zero.

$$\frac{\partial \hat{Y}}{\partial x_1} = 1.84 \times 10^{-4} - 3.596 \times 10^{-4}x_1 + 9.4 \times 10^{-5}x_2$$
$$- 9.489 \times 10^{-5}x_3 + 1.723 \times 10^{-5}x_4$$

(4.30)

$$\frac{\partial \hat{Y}}{\partial x_2} = 9.257 \times 10^{-5} - 1.8862 \times 10^{-4}x_2 + 9.4 \times 10^{-5}x_1$$
$$+ 4.633 \times 10^{-5}x_3 - 2.72 \times 10^{-5}x_4$$

(4.31)

$$\frac{\partial \hat{Y}}{\partial x_3} = 1.566 \times 10^{-5} - 2.174 \times 10^{-4}\, x_3 - 9.489 \times 10^{-5} x_1$$

$$+ 4.633 \times 10^{-5} x_2 + 2.426 \times 10^{-5} x_4$$

(4.32)

$$\frac{\partial \hat{Y}}{\partial x_4} = 1.504 \times 10^{-5} - 1.166 \times 10^{-4} x_4 + 1.723 \times 10^{-5} x_1$$

$$-2.72 \times 10^{-5} x_2 + 2.426 \times 10^{-5} x_3$$

(4.33)

$$A = \begin{pmatrix} 3.596 \times 10^{-4} & -9.4 \times 10^{-5} & 9.489 \times 10^{-5} & -1.723 \times 10^{-5} \\ 9.4 \times 10^{-5} & -1.8862 \times 10^{-4} & 4.633 \times 10^{-5} & -2.72 \times 10^{-5} \\ 9.489 \times 10^{-5} & -4.633 \times 10^{-5} & 2.174 \times 10^{-5} & -2.426 \times 10^{-5} \\ 1.723 \times 10^{-5} & -2.82 \times 10^{-5} & 2.426 \times 10^{-5} & -1.166 \times 10^{-4} \end{pmatrix}$$

$$x = \begin{bmatrix} x_1 \\ x_2 \\ x_3 \\ x_4 \end{bmatrix}$$

$$B = \begin{pmatrix} 1.84 \times 10^{-4} \\ -9.257 \times 10^{-5} \\ 1.566 \times 10^{-5} \\ -1.504 \times 10^{-5} \end{pmatrix}$$

$$Ax = B$$

(4.34)

$$x = A^{-1}B$$

(4.35)

MATLAB® program to solve Equations 4.30 through 4.33.

$$A = \begin{bmatrix} \left(3.596 \times 10^{-4}\right)\left(9.4 \times 10^{-5}\right)\left(9.489 \times 10^{-5}\right)\left(1.723 \times 10^{-5}\right); \\ \left(-9.4 \times 10^{-5}\right)\left(-1.8862 \times 10^{-4}\right)\left(-4.633 \times 10^{-5}\right)\left(-2.82 \times 10^{-5}\right); \\ \left(9.489 \times 10^{-5}\right)\left(4.633 \times 10^{-5}\right)\left(2.174 \times 10^{-5}\right)\left(2.426 \times 10^{-5}\right); \\ \left(-1.723 \times 10^{-5}\right)\left(-2.72 \times 10^{-5}\right)\left(-2.426 \times 10^{-5}\right)\left(-1.166 \times 10^{-4}\right) \end{bmatrix}$$

$$b = \left[\left(1.84 \times 10^{-4}\right); \left(-9.257 \times 10^{-5}\right); \left(1.566 \times 10^{-5}\right); \left(-1.504 \times 10^{-5}\right)\right]$$

$$x = \text{inv}(a) * b$$

Output of the program

$$x = \begin{pmatrix} 0.0 \\ 0.531 \\ 0.1902 \\ 0.0447 \end{pmatrix}$$

x_1 = coded value of optimal dextrose = 0.0
x_2 = coded value of optimal peptone = 0.531
x_3 = coded value of optimal yeast extract = 0.1902
x_4 = coded value of optimal malt extract = 0.0447

4.3.2.1.2 Simplex Contour Optimization

This is one of the easiest and simplest optimization algorithms. This method involves searching for the optimum in a given domain for experimentation. It has the flexibility of searching a space of any number of dimensions, but it can be easily visualized for a two-dimensional or a three-dimensional case. Simplex has the advantage of being able to easily compute numerically, or in a two- or three-dimensional case, it can have geometric solution.

For example, in a two-dimensional simplex problem, there are two variables that control the response of a system. First, three random points are chosen for experimentation, experiments are conducted at these points, and the responses are recorded. Now, the different points are ranked as the first, second, and the worst experimentation (F, S, and W). These three points represent a triangle in space and the mid-point of the side opposite to the point W (worst) is identified as M. The point W is the reflection of the point, M. The reflected point appears on the other side of FS is called point R. The experiment is performed at this point. Then the three new points of consideration F, S, and R are again compared, and the worst point is eliminated by reflection about the mid-point of the opposite side. This process is repeated multiple times until a specific target productivity attains or to do a specified number of simplex reflections.

There could be one obvious question. What happens if the point R happens to be worse than F and S then if we reflect this point about M we get W, which is not much of use to us? In such a case, one can take F as the best operational point or can use extension method or contraction method. In the extension method, the next point chosen after R, if R performs worse than F and S is a new point E, which is a point in the extension of R and M. Typically, R is the mid-point of EM. Similarly, using the contraction method, the new experimentation point C is along the line of W and M at the mid-point (Figure 4.4).

In a more simplified way, different movement of the basic simplex is summarized in Figure 4.5

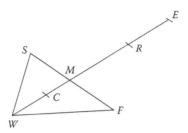

Experimental domain for
bioprocess optimization

FIGURE 4.4
Domain for search of optimal conditions.

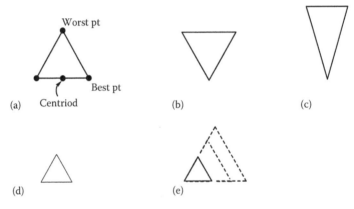

FIGURE 4.5
Various operations with the basic simplex. (a) Initial conditions for triangle, (b) reflection, (c) reflection and expansion, (d) contraction, and (e) series contraction.

Consider the following problem: A biological system, which is being optimized, has a clear domain as given below and the response contours are drawn with the corresponding values mentioned next to it. This system is defined by two individual variables A and B. Using the simplex contour method determine the response that can be achieved after five simplex operations.

Figure 4.6 is the contours for the given problem with the scale of the domain mentioned in it.

Now, using the simplex method, a more productive point is desired after five iterations. For this, three random points are on the experimentation domain (Figure 4.7).

Of these points, A and B lie on a contour of value 2.1, that is the response at point A is 2.1 units. The experimental response at point C is 2.3 units. One of the two points A or B is random. B is the worst point, which is eliminated

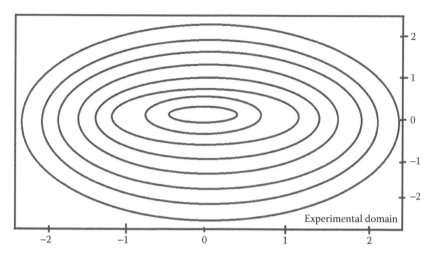

FIGURE 4.6
Contours with the domain.

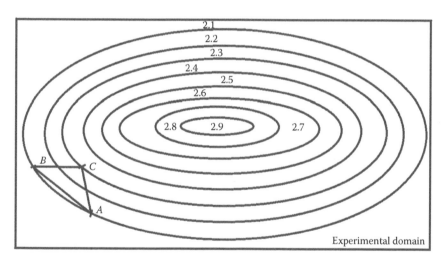

FIGURE 4.7
Progress of contour step 1.

by reflecting it about the mid-point of AC. This leads to the following change and the formation of BCD. *D* is a reflection of *B* on AC (Figure 4.8).

Now, ACD has been used for the next set of simplex operations. Similar to the previous iterations, the worst-performing point in this case is *A*. Now, *A* is eliminated to yield *E* as the new point (Figure 4.9).

The triangle with *E* is completed with line segment CD and the worst-performing point of this set, which is *D*, is then eliminated to yield triangle CEF (Figure 4.10).

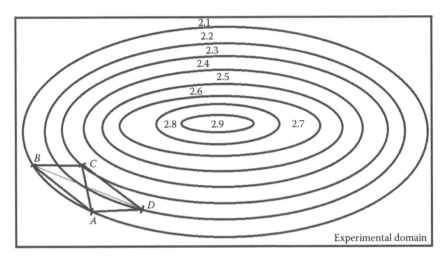

FIGURE 4.8
Progress of contour step 2.

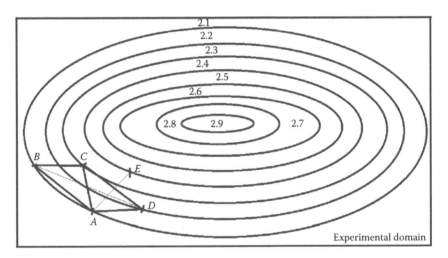

FIGURE 4.9
Progress of contour step 3.

Again, the same process is repeated and the point C being the low-performing experimental point is eliminated to have triangle EFG. In this case, total number of iterations is five. The best point among three experimental points is point G, with a response of 2.6 units (Figure 4.11).

This point is used as the result. This can be mapped onto the initially shown contour that has the values of the two axes (which represent the two variables that are being controlled/manipulated) thus giving the exact coordinates of this point (conditions of operation).

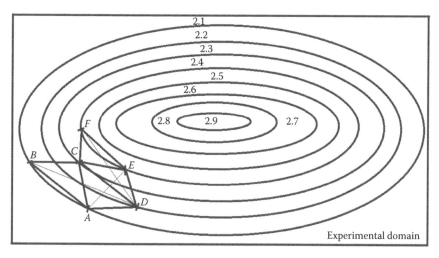

FIGURE 4.10
Progress of contour step 4.

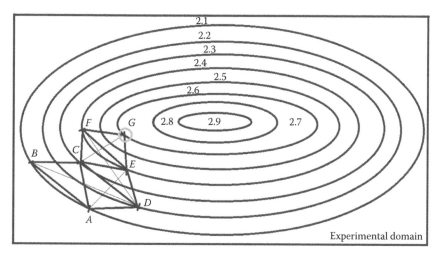

FIGURE 4.11
Progress of contour step 5.

After five iterations, the response is 2.6 units for this given biological system guided by this bioprocess when operated at point G.

In a real-life problem, there is no such defined contour. Only after experiments, the value of response is known at a given point. Therefore, a simplex guides the evolution of solutions in an improving direction.

Bruns et al.[1] have suggested five rules for the optimization by the simplex method. This is for basic simplex. They have suggested detailed methods for the modified and super modified simplexes. Readers practicing biological experiments might appreciate the basic simplex optimization. Therefore, a few similar problems in this direction will highlight the optimization of experimental variables.

Example 4.7

Figure 4.12 represents contour plot showing the effect of temperature and digestion time on the percentage purity of a biopolymer recovered using sodium dodecyl sulphate. Digestion time and temperature are the experimental variables. Use simplex method of optimization to locate the optimal conditions of the process variables. Explain the five basic rules of Bruns et al.[1] using this example.

Solution:
Five basic rules of Bruns et al.[1] are described here:

> *Rule 1*: The first simplex is determined performing a number of experiments = number of factors +1.
> First simplex is a triangle *ABC*, whose vertices *A, B, C* correspond to worst, second-worst, and best responses of the simplex.

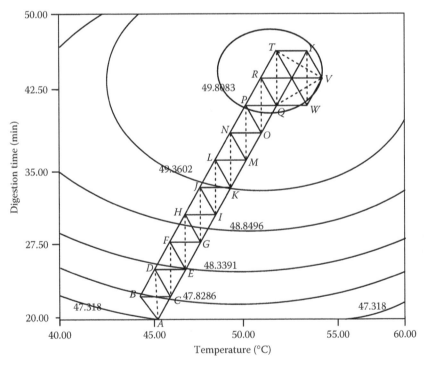

FIGURE 4.12
Isoresponse contour plot of digestion time (in minutes) and temperature (in °C).

Rule 2: The new simplex is formed rejecting the matrix corresponding to the worst vertex (here, A) and substituting it by its mirror image through the hyper-face defined the other vertices (B and C). This produces D vertex. The new simplex is BCD. The worst response is at B now. Reflect B across CD to get simplex CDE and so on.

Rule 3: When the reflected vertex has the worst response, the second-worst vertex is rejected and continued with Rule 1 and Rule 2.

Rule 4: If same vertex is maintained in $(p + 1)$ simplexes (p is the number of independent parameters; here, $p = 2$), the value of response at this vertex is measured again. If new response is lower, the vertex is rejected. If it is still high, the value is kept in the simplex.

Rule 5: If new vertex passes beyond acceptable limits for any variable that is being adjusted, an undesirable value to this vertex is attributed. Rules 2 and 3 will force the simple back to the acceptable region.

Completion: When the simplex arrives in the neighbourhood of the maximum, which is the desired value, the simplex begins to describe a circular movement around the point with highest response. The optimization process ends here.

Steps in simplex movement: In this case, there are two variables. Hence, the simplex is a triangle. The first triangle is ABC.

Reflection of a simplex vertex $\bar{A}$ above an edge BC is given by

Initial simplex ABC – 1
A: worst; B: second-worst; and C: best vertices.
Defining the vertices

$$A_x = 45 + \left(\frac{2 \times 5}{27}\right) \left(\text{scale on } x\text{-axis}; 27 \text{ mm} \equiv 5°C\right)$$

$$= 45.37$$

$$A_y = 20$$

Therefore, $\bar{A} = (45.37, 20)$.
ABC is equilateral; we can define B and C.
Take $[(B + C)/2 - \bar{A})_y] = 8$ mm, which corresponds to 2.31 min.
Therefore, $B_y = C_y = (20 + 2.31)$ min $= 22.31$ min.
Each side of equilateral $\Delta = 8$ mm $\times (2/\sqrt{3}) \approx 9.23$ mm ≈ 9 mm

$$B_x = 45.37 - (4.5)\left(\frac{5}{27}\right)$$

$$= 44.53$$

$$C_x = 45.37 + (4.57)\left(\frac{5}{27}\right) = 46.29$$

Therefore, $B = (44.53, 22.31)$ and $C = (46.29, 22.31)$.

Simplex 2: Take reflection of $\bar{A}$ around BC to get D.

$$D = B + C - \bar{A}$$
$$= \left(44.53 + 46.2 - 45.37,\ 22.31 + 22.31 - 22.31\right)$$
$$= \left(45.36,\ 24.62\right)$$

Simplex 3: Reflect B about DC to get E.

$$E = D + C - B$$
$$= \left(45.36 + 46.2 - 44.53,\ 24.62 + 22.31 - 22.31\right)$$
$$= \left(47.03,\ 24.62\right)$$

Simplex 4: Reflect C about DE to get F.

$$F = D + E - C$$
$$= \left(47.36 + 47.03 - 46.2,\ 24.62 + 24.62 - 22.31\right)$$
$$= (46.19,\ 26.93)$$

Simplex 5: Reflect D about EF to get G.

$$G = E + F - D$$
$$= \left(47.03 + 46.19 - 45.36,\ 26.93 + 24.62 - 24.62\right)$$
$$= \left(47.86,\ 26.93\right)$$

Simplex 6: Reflect E across FG to get H.

$$H = F + G - E$$
$$= \left(46.19 + 47.86 - 47.03,\ 26.93 + 26.93 - 24.62\right)$$
$$= \left(47.03,\ 29.24\right)$$

Simplex 7: Reflect F across GH to get I.

$$I = G + H - F$$
$$= \left(47.86 + 47.02 - 46.19,\ 29.24 + 26.93 - 26.93\right)$$
$$= (48.69,\ 29.24)$$

Simplex 8: Reflect G across HI to get J.

$$J = H + I - G$$
$$= (47.02 + 48.68 - 47.86, 29.24 + 29.24 - 26.93)$$
$$= (47.86, 31.55)$$

Simplex 9: Reflect H across JI to get K.

$$K = J + I - H$$
$$= (47.86 + 48.69 - 47.02, 31.55 + 29.24 - 29.24)$$
$$= (49.53, 21.55)$$

Simplex 10: Reflect I across JK to get L.

$$L = J + K - I$$
$$= (49.53 + 47.86 - 48.69, 31.55 + 31.55 - 29.24)$$
$$= (48.69, 33.7)$$

Simplex 11: Reflect J across LK to get M.

$$M = L + K - J$$
$$= (48.7 + 49.53 - 47.86, 33.7 + 31.55 - 31.55)$$
$$= (50.37, 33.7)$$

Simplex 12: Reflect K across LM to get N.

$$N = L + M - K$$
$$= (48.70 + 50.37 - 49.53, 33.7 + 33.7 - 31.55)$$
$$= (49.53, 35.85)$$

Simplex 13: Reflect L across NM to get O.

$$O = N + M - L$$
$$= (49.53 + 50.37 - 48.69, 35.85 + 33.70 - 33.7)$$
$$= (51.21, 35.85)$$

Simplex 14: Reflect M across ON to get P.

$$P = O + N - M$$
$$= \left(51.21 + 49.53 - 50.37,\ 35.85 + 35.85 - 33.70\right)$$
$$= \left(50.37,\ 38\right)$$

Simplex 15: Reflect N across OP to get Q.

$$Q = P - Q + O$$
$$= \left(50.37 + 51.21 - 49.53,\ 38.00 + 35.85 - 35.85\right)$$
$$= \left(52.05,\ 38\right)$$

Simplex 16: Reflect O across PQ to get R.

$$R = P + Q - O$$
$$= \left(50.37 + 52.05 - 51.21,\ 38 + 38 - 35.85\right)$$
$$= \left(51.21,\ 40.15\right)$$

Simplex 17: Reflect P across QR to get S.

$$S = Q + R - P$$
$$= \left(52.05 + 51.21 - 50.37,\ 40.15 + 38 - 38\right)$$
$$= \left(52.89,\ 40.15\right)$$

Simplex 18: Reflect Q across RS to get T.

$$T = R + S - Q$$
$$= \left(51.27 + 52.89 - 52.05,\ 40.15 + 40.15 - 38\right)$$
$$= \left(52.05,\ 42.30\right)$$

Simplex 19: Reflect R across ST to get U.

$$U = S + T - R$$
$$= \left(52.05 + 52.89 - 51.21,\ 42.30 + 40.15 - 40.15\right)$$
$$= \left(53.73,\ 42.30\right)$$

Simplex 20: Reflect T across US to get V.

$$V = U + S - T$$

$$= \left(52.89 + 53.73 - 52.05, \; 42.30 - 42.30 + 40.15\right)$$

$$= \left(54.57, \; 40.17\right)$$

Simplex 21: Reflect U across VS to get W.

$$W = V + S - U$$

$$= \left(54.57 + 52.89 - 53.73, \; 40.15 + 40.15 - 42.30\right)$$

$$= \left(53.73, \; 38\right)$$

Simplex 22: Reflect V across SW to get V'.

$$V' = S + W - V$$

$$= \left(53.73 - 54.57 + 52.89, \; 38 - 40.15 + 40.15\right)$$

$$= \left(52.05, \; 38\right) \text{ which is the same as } Q$$

$$V' = Q = \left(52.05, \; 38\right)$$

Hence, the process comes to a completion. Here, the simplex revolves around S after reaching optimum. S is the required point.

$$S = \left(52.89, \; 40.15\right)$$

Optimum temperature $= 52.89°C$
Optimum digestion time $= 40.15$ min.

Example 4.8

Figure 4.13 shows the contour plots of temperature versus sodium dodecylsulphate (SDS) concentration on percentage purity of a biopolymer recovered using SDS method. Use the simplex method for optimization of process variables.

Solution:

For initial simplex $\triangle ABC$, that is, simplex 1, vertex A corresponds to the worst response.
Scale on x-axis: 26 mm $\equiv 5°C$ and

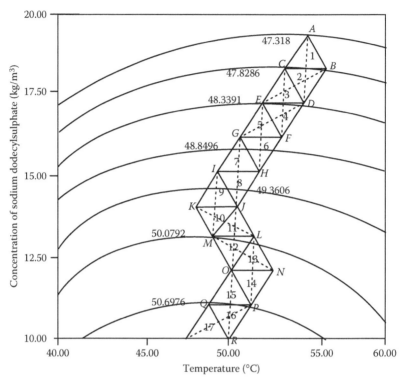

FIGURE 4.13
Isoresponse contour plots of concentration of sodium dodecylsulphate and temperature. Response is the percent purity of the product.

For y-axis: 26 mm $\equiv$ 2.5 kg/m³.

$$A_x = 50 + \left(\frac{18 \times 5}{26} \right) = 53.46$$

$$A_y = 17.5 + \left(\frac{2.5 \times 12}{26} \right) = 18.6$$

Hence, the coordinate for $\bar{A}$ = (53.46, 18.6).
Length of side = 1.13 cm

$$B_x = 53.46 + \left(\frac{6 \times 5}{26} \right) = 54.61$$

$$C_x = 53.46 - \frac{6 \times 5}{26} = 52.31$$

$$C_y = B_y = 18.6 - \left(\frac{2.5 \times 10}{26} \right) = 17.64$$

Therefore, $B = (54.61, 17.64)$ and $C = (52.31, 17.64)$.
Throughout the movement of the simplex, Rule 1 of Bruns et al.[1] is applicable.

Simplex 2: Reflect $\bar{A}$ across BC to get E.

$$D = B + C - \bar{A}$$
$$= \left(54.61 + 52.31 - 53.46,\ 17.64 + 17.64 - 18.6 \right)$$
$$= \left(53.46,\ 16.68 \right)$$

Simplex 3: Reflect B across CD to get E.

$$E = D + C - B$$
$$= \left(53.46 + 52.31 - 54.61,\ 16.68 + 17.64 - 17.64 \right)$$
$$= \left(51.16,\ 16.68 \right)$$

Simplex 4: Reflect C across DE to get F.

$$F = D + E - C$$
$$= (51.16 + 53.46 - 52.31,\ 16.68 + 16.68 - 17.64)$$
$$= \left(52.75,\ 15.72 \right)$$

Simplex 5: Reflect D across EF to get G.

$$G = E + F - D$$
$$= \left(51.16 + 52.31 - 53.46,\ 16.68 + 15.72 - 16.68 \right)$$
$$= \left(50.01,\ 15.72 \right)$$

Simplex 6: Reflect E across GF to get H.

$$H = G + F - E$$
$$= \left(50.01 + 52.31 - 51.16,\ 15.72 + 15.72 - 16.68 \right)$$
$$= \left(51.16,\ 14.76 \right)$$

Simplex 7: Reflect F across GH to get I.

$$I = G + H - F$$
$$= (50.01 + 51.16 - 52.31,\ 15.72 + 14.76 - 15.72)$$
$$= \left(48.86,\ 14.76 \right)$$

Simplex 8: Reflect G across IH to get J.

$$J = I + H - G$$
$$= (48.86 + 51.16 - 50.01, \ 14.76 + 14.76 - 15.72)$$
$$= (59.01, \ 13.8)$$

Simplex 9: Reflect H across IJ to get K.

$$K = I + J - H$$
$$= (48.86 + 50.01 - 51.16, \ 14.76 + 13.8 - 14.76)$$
$$= (47.71, \ 13.8)$$

Simplex 10: Reflect I across KJ to get L.

$$L = K + J - I$$
$$= (47.71 + 50.01 - 48.86, \ 13.5 + 13.8 - 14.76)$$
$$= (48.86, \ 12.84)$$

Simplex 11: Reflect K across LJ to get M.

$$M = L + J - K$$
$$= (48.86 + 50.01 - 47.71, \ 12.84 + 13.8 - 13.8)$$
$$= (51.16, \ 12.84)$$

Simplex 12: Reflect J across LM to get N.

$$N = L + M - J$$
$$= (51.16 + 48.86 - 50.01, \ 12.84 + 12.84 - 13.8)$$
$$= (50.01, \ 11.85)$$

Simplex 13: Reflect L across NM to get O.

$$O = N + M - L$$
$$= (50.01 + 51.16 - 48.86, \ 11.85 + 12.84 - 12.84)$$
$$= (52.31, \ 11.85)$$

Simplex 14: Reflect M across ON to get P.

$$P = O + N - M$$
$$= (52.31 + 50.01 - 51.16, \ 11.85 + 11.85 - 12.84)$$
$$= (51.16, \ 10.86)$$

Simplex 15: Reflect O across PN to get Q.

$$Q = P + N - O$$

$$= (51.16 + 50.01 - 52.31, 10.86 + 11.85 - 11.85)$$

$$= (48.86,\ 10.86)$$

Simplex 16: Reflect N across PQ to get R.

$$R = P + Q - N$$

$$= (51.16 + 48.6 - 50.01, 10.86 + 10.86 - 11.85)$$

$$= (50.01,\ 10)$$

Simplex 17: Reflect P across QR to get S.

$$S = Q + R - P$$

$$= (48.86 + 50.01 - 51.16, 10.86 + 10 - 10.86)$$

$$= (47.71,\ 10)$$

The simplex cannot be extended further, although the optimum values could not be achieved so far.

It is safe to say that $S = (47.71, 10)$ and $R = (50.01, 10)$ are close to optimal conditions, that is, $10\ \text{kg/m}^3$; 47.71 and 50.01°C.

4.3.2.2 Second-Order Surface for Multi-Response

Let one consider the multi-response experiments and its analysis.

In multi-response experiments, a number of responses are measured simultaneously for each setting of group of input variables. For example, the production of pectinase, esterase, and carboxymethyl cellulase is coupled with the cell growth. In this experiment, there are four responses, which are influenced by apparent independent variables: pH, temperature, and agitation. One set of responses is obtained for each setting of independent variables. Table 4.27 gives an example of setting of the variables for the optimization of biological parameters. Table 4.28 gives the responses for enzyme activity and cell concentration for corresponding setting of apparent independent variables in Table 4.27.

The coding of the variables is described for better concept.

x_1 is the coded value of the variable X_1 (slant age, h) = [X_1 (actual slant age, h) – 47]/17,

x_2 is the coded value of the variable X_2 (inoculum age, h) = [(X_2 (actual inoculum age, h) – 12]/4

TABLE 4.27

Optimization of the Biological Parameters

Run Number	$x_1 (\equiv X_1)$	$x_2 (\equiv X_2)$	$x_3 (\equiv X_3)$ 1·10^8
1	−1 (≡ 30)	−1 (≡ 8)	−1 (≡ 1)
2	1 (≡ 64)	−1 (≡ 8)	−1 (≡ 1)
3	−1 (≡ 30)	1 (≡ 16)	−1 (≡ 1)
4	1 (≡ 64)	1 (≡ 16)	−1 (≡ 1)
5	−1 (≡ 30)	−1 (≡ 8)	1 (≡ 2)
6	1 (≡ 64)	−1 (≡ 8)	1 (≡ 2)
7	−1 (≡ 30)	1 (≡ 16)	1 (≡ 2)
8	1 (≡ 64)	1 (≡ 16)	1 (≡ 2)
9	−1.682 (≡18.41)	0 (≡ 12)	0 (≡ 1.5)
10	1.682(≡75.59)	0 (≡ 12)	0 (≡ 1.5)
11	0 (≡47)	−1.682 (≡ 5.27)	0 (≡ 1.5)
12	0 (≡ 47)	1.682 (≡18.73)	0 (≡ 1.5)
13	0 (≡ 47)	0 (≡ 12)	−1.682(≡0.659)
14	0 (≡ 47)	0(≡ 12)	1.682 (≡ 2.341)
15	0 (≡ 47)	0 (≡ 12)	0 (≡ 1.5)
16	0 (≡ 47)	0 (≡ 12)	0 (≡ 1.5)
17	0 (≡ 47)	0 (≡ 12)	0 (≡ 1.5)
18	0 (≡ 47)	0 (≡ 12)	0 (≡ 1.5)
19	0 (≡ 47)	0 (≡ 12)	0 (≡ 1.5)
20	0 (≡ 47)	0 (≡ 12)	0 (≡ 1.5)

Note: The actual and coded values of variables are given in and outside of parenthesis, respectively.

x_3 is the coded value of the variable X_3 (number of cells/ml) = [(X_3 (actual number of cells/ml) − 1.5 × 10^8)]/0.5 × 10^8.

In this case, any design criterion should be based on perceiving the responses as a group rather than as individualized entities. Hence, in a multi-response situation, therefore, the choice of a design should be based on a criterion that incorporates measures of efficiency pertaining to all of the responses. It is rarely the case where all response variables achieve their respective optima at the same set of conditions. One might consider superimposing contours of all response variables and then pinpoint a region where conditions can be *near* optimal for all the responses.

The components in such study are the following.

- Estimation of parameters
- Design and analysis

TABLE 4.28

Cell Concentration and Representative Enzyme Activity Corresponding to the Setting of Table 4.27

Run No.	Slant Age, h (x_1)	Inoculum Age h (x_2)	Number of Cells, (Cells/ml (x_3) 1×10^8)	Enzyme Activity (U)	Cell Concentration (kg/m³)
2	1	−1	−1	1.381×10^{-3}	6.9
3	−1	1	−1	1.351×10^{-3}	6.6
4	1	1	−1	1.411×10^{-3}	6.5
5	−1	−1	1	1.593×10^{-3}	6.6
6	1	−1	1	1.114×10^{-3}	6.7
7	−1	1	1	1.419×10^{-3}	6.9
8	1	1	1	1.502×10^{-3}	6.9
9	−1.682	0	0	1.486×10^{-3}	6.5
10	1.682	0	0	1.409×10^{-3}	6.9
11	0	−1.682	0	1.565×10^{-3}	7.1
12	0	1.682	0	1.529×10^{-3}	6.6
13	0	0	−1.682	1.371×10^{-3}	6.6
14	0	0	1.682	1.354×10^{-3}	6.7
15	0	0	0	1.853×10^{-3}	6.5
16	0	0	0	1.824×10^{-3}	6.4
17	0	0	0	2.019×10^{-3}	6.1
18	0	0	0	1.641×10^{-3}	6.2
19	0	0	0	1.954×10^{-3}	5.5
20	0	0	0	1.798×10^{-3}	5.6

- Testing of lack-of-fit
- Simultaneous optimization of several responses variables

Two different models are linear and generalized multi-response models.[4] For the optimization of multi-response function, Khuri and Cornell[4] suggested optimization using the desirability function and the generalized distance function approaches. The desirability function approach is easier as the overall desirability is a continuous function of the input variables.[9,10] Improper desirability might produce incorrect decision. The generalized distance function approach does not consider either heterogeneity of variances of the responses or interrelationship among the responses. In this case, the generalized distance function approach is explained with the proper example.

4.3.2.2.1 Generalized Distance Function Approach

Khuri and Cornell[4] described the elaborate theory of the generalized distance function approach.

The regression equations for response variables are used in distance function ($\delta[Y, \varphi]$). This equation is minimized using MATLAB. Simultaneous optimum values are substituted in the regression equation. If the value of δ is very close to zero, it indicates that simultaneous optimal conditions approach *ideal* optimal conditions.

Example 4.9

A fungus synthesizes an enzyme under optimal conditions of independent variables, obtained from single response analysis. The optimal enzyme activity is 0.9 U. Experiments followed the CCD as per the matrix given in Table 4.29. The response is expressed by second-order model equation, having 10 parameters. The experimenter is interested to analyse the results using generalized distance function approach of multi-response analysis. Determine the boundaries of rectangular confidence region.

TABLE 4.29

Design Matrix for Enzyme Synthesis Using
Central Composite Design Approach

x_1	x_2	x_3
C-Source	N-Source	Minerals
−1	−1	−1
+1	−1	−1
−1	+1	−1
+1	+1	−1
−1	−1	+1
+1	−1	+1
−1	+1	+1
+1	+1	+1
−1.68	0	0
+1.68	0	0
0	−1.68	0
0	+1.68	0
0	0	−1.68
0	0	+1.68
0	0	0
0	0	0
0	0	0
0	0	0
0	0	0
0	0	0

Solution:

This is a matrix of order 20×3.

$$
X_0 =
\begin{bmatrix}
-1 & -1 & -1 \\
+1 & -1 & -1 \\
-1 & +1 & -1 \\
+1 & +1 & -1 \\
-1 & -1 & +1 \\
+1 & -1 & +1 \\
-1 & +1 & +1 \\
+1 & +1 & +1 \\
-1.68 & 0 & 0 \\
+1.68 & 0 & 0 \\
0 & -1.68 & 0 \\
0 & +1.68 & 0 \\
0 & 0 & -1.68 \\
0 & 0 & +1.68 \\
0 & 0 & 0 \\
0 & 0 & 0 \\
0 & 0 & 0 \\
0 & 0 & 0 \\
0 & 0 & 0 \\
0 & 0 & 0
\end{bmatrix}
$$

X_0^T is a matrix of order 3×20.

$$
(X_0^T X_0)^{-1} =
\begin{bmatrix}
0.088 & 0 & 0 \\
0 & 0.088 & 0 \\
0 & 0 & 0.088
\end{bmatrix}
$$

$$
z(\psi_i) =
\begin{pmatrix}
-0.7162 \\
0.6165 \\
0.245
\end{pmatrix}
$$

$$z^T(\psi_i) = (-0.7162 \quad 0.6165 \quad 0.245)$$

$$z^T(\psi_i)(X_0^T X_0)^{-1} = (-0.063 \quad 0.0542 \quad 0.2156)$$

$$z^T(\psi_i)(X_0^T X_0)^{-1}z(\psi_i) = (-0.063 \quad 0.0542 \quad 0.2156)\begin{pmatrix} -0.7162 \\ 0.6165 \\ 0.245 \end{pmatrix}$$

$$= 0.0838$$

$$g_l(X_0, \psi_i) = [z^T(\psi_i)(X_0^T X_0)^{-1}z(\psi_i)]^{1/2} = (0.0838)^{1/2} = 0.2895$$

$$\kappa_{1i} = \varphi_i - g_i(X_0, \psi_i)(MS_i t_{\alpha/2,(N-P)})^{1/2}$$

In this case,

$i = 1$
N = number of experiments = 20
P = number of parameters = 10

$$N - P = 20 - 10 = 10$$

Confidence level, $\alpha = 0.11$ and $\alpha/2 = 0.055$.
   From statistical table, $t_{\alpha/2,N-P} = t_{0.055,10} = 1.81$.
   Root mean square error = 0.00312

$$\varphi_i = 0.90$$

$$\kappa_{1_i} = 0.90 - [(0.2895)(0.00312 \times 1.81)]^{1/2}$$

$$= 0$$

$$\kappa_{2_i} = \varphi_i + g_i(X_0, \psi_i)(MS_i t_{\alpha/2,(N-P)})^{1/2}$$

$$\kappa_{2_i} = 0.90 + [(0.2895)(0.00312 \times 1.81)]^{1/2}$$

$$= 0$$

Exercises

4.1 Calculate *F*-statistic and R^2 for the following CCD plan, given in Table 4.1.1. Construct the ANOVA table. Assume $F_{\text{tabulated value}}$ at a particular probability = 12.9. Interpret the result.

4.2 (a) Write the applications of Plackett-Burman design. Calculate the effect and confidence level. (b) State and calculate the elements in a typical ANOVA table. Following information is available.

Y_i is the observed response

N is the number of observation

$\tilde{Y}$ is the predicted response

P is the probability

β_i are the coefficients in the polynomial assumed for

H_0 is the null hypothesis

H_{alt} is the alternate hypothesis

F is the *F*-statistic

α is the level of significance

4.3 Figure 4.3.1 is an isoresponse contour plot with proper simplex movement. Find the optimal conditions.

4.4 Figure 4.4.1 is an isoresponse contour plot at different combinations of the two variables for enzyme activity (U). The optimal response is nearer to 2.2308 U. Use the simplex method having initial conditions

TABLE 4.1.1

CCD Design Plan

Run Number	Temperature (x_1)	Initial pH (x_2)	Experimental Response (Y_i)	Predicted Response
1	−1	−1	0.3367	0.3320
2	1	−1	0.3448	0.3440
3	−1	1	0.2587	0.2607
4	1	1	0.2728	0.2788
5	−1	0	0.3536	0.3563
6	1	0	0.3766	0.3714
7	0	−1	0.3571	0.3626
8	0	1	0.3023	0.2943
9	0	0	0.3914	0.3884
10	0	0	0.3817	0.3884
11	0	0	0.3892	0.3884
12	0	0	0.3921	0.3884
13	0	0	0.3852	0.3884

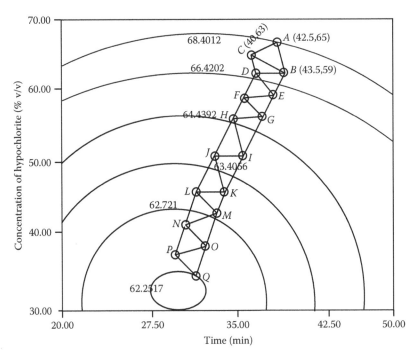

FIGURE 4.3.1
Hypochlorite concentrations versus time for digestion with reference to a biopolymer extraction from cells.

of pH and temperature described in Table 4.4.1. Show the coordinates at different condition of movement of the simplex to attain optimal conditions of pH and temperature corresponding to the response of 2.2308 U.

4.5 Isoresponse contour plots at different levels of digestion time of cells and the pH of the digestion mixture on the percentage recovery of a biopolymeris given in Figure 4.5.1. Determine the optimal conditions of the variables by simplex method.

4.6 Isoresponse contour plots at different levels of hypochlorite percentage and the pH of the digestion mixture on the percentage recovery of a biopolymer is given in Figure 4.6.1. Determine the optimal conditions of the variables by simplex method.

4.7 Isoresponse contour plots at different levels of digestion time of cells and the pH of the digestion mixture on the percentage recovery of a biopolymer using SDS method is given in Figure 4.7.1. Determine the optimal conditions of the variables by simplex method. Interpret similar results obtained from the hypochlorite extraction method.

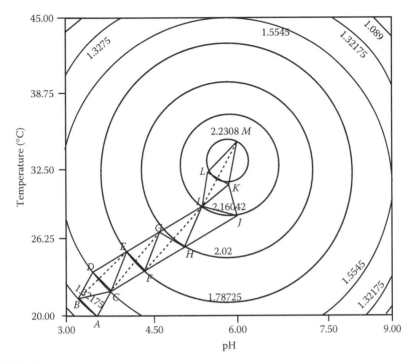

FIGURE 4.4.1
pH versus temperature isoresponse contour plots for esterase production.

TABLE 4.4.1

Temperature and pH Conditions for Initial Simplex

pH	Temperature (°C)
3.75	20
3.4	22
4.1	22

4.8 Isoresponse contour plots at different levels of digestion temperature and the pH of the digestion mixture on the percentage recovery
of a biopolymer using SDS method is given in Figure 4.8.1. Determine
the optimal conditions of the variables by simplex method.

4.9 Isoresponse contour plots at different levels of digestion time temperature and the pH of the digestion mixture on the percentage
purity of a biopolymer is given in Figure 4.9.1. Determine the optimal conditions of the variables by simplex method.

4.10 Isoresponse contour plots at different levels of temperature and the
hypochlorite concentration on the purity of biopolymer recovered by
hypochlorite digestion method is given in Figure 4.10.1. Determine

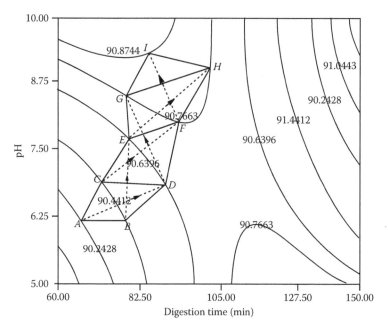

FIGURE 4.5.1
Contour plots for pH versus digestion time.

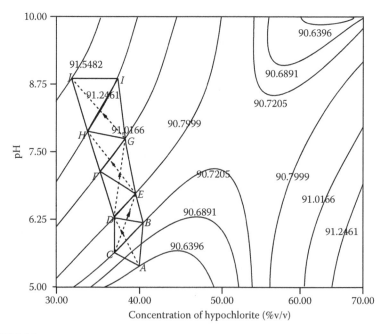

FIGURE 4.6.1
Isoresponse contour plots for biopolymer recovery.

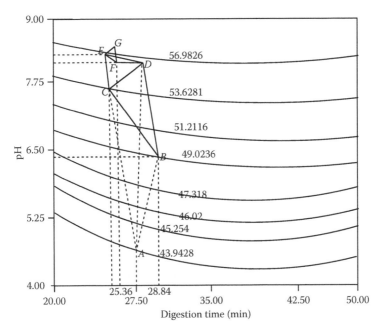

FIGURE 4.7.1
Contour plots for biopolymer recovery obtained by SDS method.

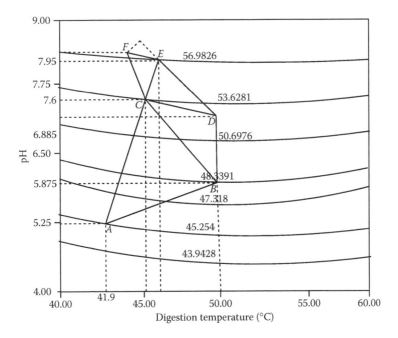

FIGURE 4.8.1
Biopolymer extraction using SDS method.

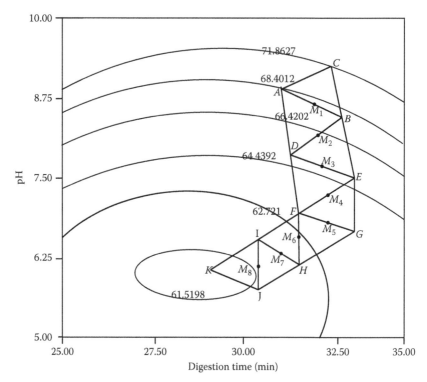

FIGURE 4.9.1
Contour plots for purity of biopolymer.

the optimal conditions of the variables by simplex method. Which is the rule of Bruns et al.[1] suggested in this case?

4.11 The central-composite experimental design was suggested for the optimization of pH (initial) and temperature for the production of carboxymethylcellulase synthesized by a fungal system. The range of pH (initial) and temperature of fermentation was between 5 and 7, and between 20°C and 30°C, respectively. Answer the following:

a. Code pH (initial) and temperature of fermentation as per CCD.

b. Write the matrix for experimental purposes.

c. How many factorial points are indicated in the matrix? What are they?

d. How will you decide the number of centre points? Express your answer based on total number of design points.

e. What are the criteria for rotatability?

4.12 The response equation for an enzyme synthesis in submerged culture is expressed by the following Equation 4.33.

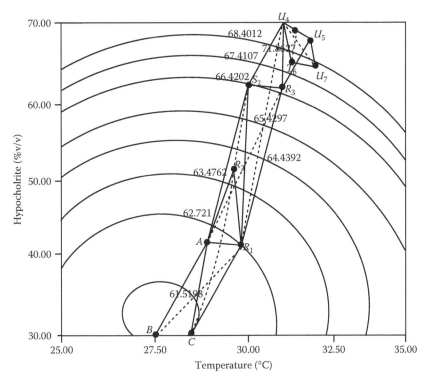

FIGURE 4.10.1
Contours plots of purity of biopolymer under different conditions.

$$\tilde{Y}_{laccase} = 748.74 + 90.54x_1 + 30.02x_2 + 9.20\,x_3 - 82.21x_1x_2$$
$$-69.24x_1x_3 + 2.43x_2x_3 + 4.45x_1^2 - 31.95x_2^2 + 12.75x_3^2$$

(4.33)

where:

$\tilde{Y}_{laccase}$ is the predicted response

x_1, x_2, and x_3 are coded values of the variables, respectively

Calculate optimum values of the variables.

4.13 For the following matrix (Table 4.13.1), suggest the experimental design plan. Suggest the model equation and optimize the variables.

4.14 Analyze the design matrix (Table 4.14.1) for ANOVA, F-statistic, R^2, and $R^2_{adjusted}$. Calculate optimal values of the variables and interpret the results.

4.15 Based on the available data in Table 4.15.1, draw possible contour plots.

TABLE 4.13.1

Experimental Matrix

Run No.	x_1	x_2	x_3	x_4	Tartaric Acid (kg/m³)	
					Experimental	Predicted
1	−1	−1	−1	−1	4.0	4.36
2	1	−1	−1	−1	5.0	5.35
3	−1	1	−1	−1	4.5	4.59
4	1	1	−1	−1	5.7	5.86
5	−1	−1	1	−1	5.2	5.24
6	1	−1	1	−1	5.9	5.76
7	−1	1	1	−1	3.5	3.99
8	1	1	1	−1	4.0	4.79
9	−1	−1	−1	1	4.2	4.14
10	1	−1	−1	1	5.6	5.06
11	−1	1	−1	1	5.1	5.09
12	1	1	−1	1	5.6	6.29
13	−1	−1	1	1	6.2	6.09
14	1	−1	1	1	6.3	6.54
15	−1	1	1	1	5.6	5.57
16	1	1	1	1	6.7	6.29
17	−2	0	0	0	4.4	4.05
18	2	0	0	0	6.0	5.77
19	0	−2	0	0	5.7	5.92
20	0	2	0	0	6.6	5.9
21	0	0	−2	0	4.4	3.97
22	0	0	2	0	5.0	4.85
23	0	0	0	−2	6.2	5.37
24	0	0	0	2	6.3	6.65
25	0	0	0	0	6.6	6.23
26	0	0	0	0	6.0	6.23
27	0	0	0	0	6.4	6.23
28	0	0	0	0	6.7	6.23
29	0	0	0	0	6.1	6.23
30	0	0	0	0	5.9	6.23

TABLE 4.14.1

Experimental Matrix

Run No.	X_1	X_2	X_3	X_4	Experimental Enzyme Activity, (U)		Predicted Enzyme Activity, (U)
					Set 1	Set 2	
1	2.5	0.7	0.5	0.45	0.0702	0.0650	0.0572
2	7.5	0.7	0.5	0.15	0.1000	0.1156	0.1018
3	2.5	2.1	0.5	0.15	0.0732	0.1256	0.0908
4	7.5	2.1	0.5	0.45	0.0570	0.0554	0.0729
5	2.5	0.7	1.5	0.15	0.0570	0.0554	0.0545
6	7.5	0.7	1.5	0.45	0.0572	0.0580	0.0631
7	2.5	2.1	1.5	0.45	0.0376	0.0420	0.047
8	7.5	2.1	1.5	0.15	0.0558	0.0546	0.0626
9	5.0	1.4	1.0	0.3	0.0566	0.0566	0.0636
10	5.0	1.4	1.0	0.3	0.0566	0.0606	0.0636
11	2.5	0.7	0.5	0.15	0.0814	0.0950	0.0784
12	7.5	0.7	0.5	0.45	0.0764	0.0844	0.0725
13	2.5	2.1	0.5	0.45	0.0612	0.0616	0.0561
14	7.5	2.1	0.5	0.15	0.1044	0.1112	0.1014
15	2.5	0.7	1.5	0.15	0.0450	0.0430	0.0461
16	7.5	0.7	1.5	0.15	0.0616	0.0732	0.0673
17	2.5	2.1	1.5	0.15	0.0520	0.0392	0.0504
18	7.5	2.1	1.5	0.45	0.0381	0.0428	0.0469
19	5.0	1.4	1.0	0.3	0.0468	0.0476	0.0606
20	5.0	1.4	1.0	0.3	0.0610	0.0610	0.0606
21	0.0	1.4	1.0	0.3	0.0118	0.0130	0.0206
22	10	1.4	1.0	0.3	0.0530	0.0474	0.0483
23	5.0	0	1.0	0.3	0.0404	0.0308	0.0461
24	5.0	2.8	1.0	0.3	0.0502	0.0430	0.0423
25	5.0	1.4	0	0.3	0.0532	0.0444	0.0725
26	5.0	1.4	2.0	0.3	0.0402	0.0402	0.0277
27	5.0	1.4	1.0	0	0.0448	0.0688	0.0637
28	5.0	1.4	1.0	0.6	0.0548	0.0294	0.0267
29	5.0	1.4	1.0	0.3	0.0404	0.0404	0.0383
30	5.0	1.4	1.0	0.3	0.0608	0.0620	0.0383

TABLE 4.15.1

Experimental Data as per Design Matrix

Run No.	Slant Age, h (x_1)	Inoculum Age, h (x_2)	Number of Cells, Cells/ml (x_3) $1 \cdot 10^8$	Experimental Activity (U/mg Dry Cell Weight)		Predicted Activity (U/mg Dry Cell Weight)
				Set 1	Set II	
1	−1	−1	−1	1.692×10^{-3}	1.786×10^{-3}	1.708×10^{-3}
2	1	−1	−1	1.381×10^{-3}	1.315×10^{-3}	1.334×10^{-3}
3	−1	1	−1	1.351×10^{-3}	1.427×10^{-3}	1.33×10^{-3}
4	1	1	−1	1.411×10^{-3}	1.341×10^{-3}	1.414×10^{-3}
5	−1	−1	1	1.593×10^{-3}	1.611×10^{-3}	1.547×10^{-3}
6	1	−1	1	1.114×10^{-3}	1.114×10^{-3}	1.158×10^{-3}
7	−1	1	1	1.419×10^{-3}	1.493×10^{-3}	1.454×10^{-3}
8	1	1	1	1.502×10^{-3}	1.514×10^{-3}	1.523×10^{-3}
9	−1.682	0	0	1.486×10^{-3}	1.522×10^{-3}	1.583×10^{-3}
10	1.682	0	0	1.409×10^{-3}	1.357×10^{-3}	1.326×10^{-3}
11	0	−1.682	0	1.565×10^{-3}	1.539×10^{-3}	1.578×10^{-3}
12	0	1.682	0	1.529×10^{-3}	1.611×10^{-3}	1.567×10^{-3}
13	0	0	−1.682	1.371×10^{-3}	1.309×10^{-3}	1.372×10^{-3}
14	0	0	1.682	1.354×10^{-3}	1.318×10^{-3}	1.327×10^{-3}
15	0	0	0	1.853×10^{-3}	1.955×10^{-3}	1.879×10^{-3}
16	0	0	0	1.824×10^{-3}	1.872×10^{-3}	1.879×10^{-3}
17	0	0	0	2.019×10^{-3}	2.109×10^{-3}	1.879×10^{-3}
18	0	0	0	1.641×10^{-3}	1.633×10^{-3}	1.879×10^{-3}
19	0	0	0	1.954×10^{-3}	2.004×10^{-3}	1.879×10^{-3}
20	0	0	0	1.798×10^{-3}	1.894×10^{-3}	1.879×10^{-3}

References

1. Bruns RE, Scarminio IS, and de Barros Neto B (Eds.), *Statistical Design – Chemometrics*, Elsevier, Amsterdam, the Netherlands, 2006.
2. Montgomery DC (Ed.), *Design and Analysis of Experiments*, 7th edition, Wiley, New Delhi, India, 2009.
3. Myers RH and Montgomery DC, *Response Surface Methodology*, 2nd edition, John Wiley & Sons, New York, 2002.
4. Khuri AI and Cornell JA, *Response Surfaces Designs and Analysis*, Marcel Dekker, New York, 1987.
5. Plackett RL and Burman JP, *The Design of Optimum Multifactorial Experiments*, *Biometrika*, **33**, 305–325, 1946.
6. Rosenbrock HH, An automatic method for finding the greatest or least value of a function, *Computer Journal*, **3**, 175–184, 1960.
7. Box MJ, A new method of constrained optimization and a comparison with other methods, *Computer Journal*, **8**, 42–52, 1965.

8. Kuester JL and Mize JH, *Optimization Techniques with FORTRAN*, McGraw-Hill, New York, 1973.
9. Harrington EC, The desirability function, *Industrial Quality Control*, **21**, 494–498, 1965.
10. Derringer G and Suich R, Simultaneous optimization of several response variables, *Journal of Quality Technology*, **12**, 214–219, 1980.
11. Khuri AI and Conlon M, Simultaneous optimization of multiple responses represented by polynomial regression function, *Technometrics*, **23**, 363–375, 1981.

Further Reading

1. Amutha Devi A, Studies on 3-hydroxy-3-methyl glutaryl co-enzyme A-reductase from *Saccharomyces cerevisiae*, PhD Thesis, Indian Institute of Technology, Madras, India, 2006.
2. Dasu VV, Studies on production of griseofulvin by *Penicillium griseofulvum*, PhD Thesis, Indian Institute of Technology, Madras, India, 1999.
3. Gowrishankar BS, Studies on production of esterase from *Saccharomyces cerevisiae*, PhD Thesis, Indian Institute of Technology, Madras, India, 2006.
4. Kapat A, Synthiesis of extra-celluar chitinase by *Trichoderma harzianum* and characterization of the enzyme, PhD Thesis, Indian Institute of Technology, Madras, India, 1999.
5. Naidu GSN, Studies on behavior and production of extracellular pectinases from *Aspergillus niger*, PhD Thesis, Indian Institute of Technology, Madras, India, 1999.
6. Nair SR, Studies on synthesis of pectolytic enzymes by *Apergilllus niger*, PhD Thesis, Indian Institute of Technology, Madras, India, 1999.
7. Prasanna GL, Studies on electrotrans-formation of *Saccharomyces cerevisiae* and elctrofusion of *Saccharomyces cerevisiae* and *Trichoderma reesei*, PhD Thesis, Indian Institute of Technology, Madras, India, 1999.
8. Rao DS, Studies on comparative analysis of gluconic acid fermentation using cane molasses by free cells and immobilized whole cells of *Aspergillus niger*, PhD Thesis, Indian Institute of Technology, Madras, India, 1994.
9. Snehalatha Ch, Studies on polyhydroxybutyrate from *Alcaligens sp.*, M. S. Thesis, Indian Institute of Technology, Madras, India, 2004.
10. Srinivas R, Studies on intergeneric fusants of *Trichoderma reesei/Saccharomyces cerevisiae*, PhD Thesis, Indian Institute of Technology, Madras, India, 1999.
11. Théodore K, Studies on optimization of β-1, 3-glucanase production by *Trichoderma harzianum* NCIM 1185, PhD Thesis, Indian Institute of Technology, Madras, India, 1995.

5

Evolutionary Operation Programmes

OBJECTIVE: Evolutionary operation programme involves simple statistical calculations. Optimization is an experiment-driven one. Varieties of variables at different levels are considered as a geometric figure. This is a technique for any user without strong foundation of mathematics for exploitation of biological systems. This can be efficiently implemented for industrial biological process.

5.1 Introduction

For the improvement and modification of ongoing process, change in factors or variables of different nature will not affect the process objective. This is possible based on the concept of Darwin's theory of evolution. This method is evolutionary operation programmes (EVOP). Box[1] (1954) introduced the first EVOP method for process improvement in the industrial scale. The change in levels of process variables is small to avoid appreciable changes in the desired product characteristics. In biological processes, some information is available to use EVOP for selecting process variables.[2,3]

5.1.1 Characteristics

EVOP has the following characteristics:

- Reduced trial-and-error approach with proper objectives for finding the preferred conditions
- Multi-variable sequential search method for a decision about the process
- Desired pre-requisite information on initial settings of process variables
- Enough replication to have an idea of some pattern in the response

Examples

Following is the selected varieties of EVOP designs:

Single-factor EVOP design: If x_1 is the current level of factor used in the process, experiments will be conducted at $(x_1 - \delta)$, x_1, and $(x_1 + \delta)$ levels (Figure 5.1). This change is within the permissible limit to have satisfactory product. Evaluation of the quality of the process is done in those three levels. Decision is to select the condition producing higher quality of product. The process will continue in the desired direction until the highest product quality is achieved.

Two-factor EVOP design: This example is mentioned by both Box[1] and Bruns et al.[4] Any number of variables can be studied by the EVOP method. In these cases, a 2^2-design is employed usually for EVOP study. Figure 5.2 explains the concept of a two-variable EVOP study.

5.1.2 Why Is It Important?

EVOP method is a multi-variable search technique. The desired product quality is maintained without any interruption in the production of the existing process by proper statistical analysis of the experimental response.

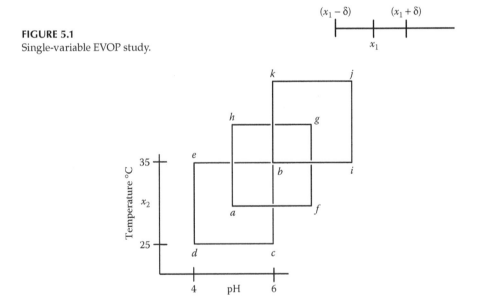

FIGURE 5.1
Single-variable EVOP study.

FIGURE 5.2
Progress of EVOP experimental design: *a* – initial condition; *b* – good condition; *d* – bad condition; *c* and *e* – intermediate conditions in the first stage of trial runs; *g* – better condition; *j* – better than of, so on; (*f, h*) and (*i, k*) are the combinations of intermediate conditions in the second and third stages of trial runs, respectively.

5.1.3 When Can We Use the EVOP Method?

The EVOP method replaces the static operation by a continuous and systematic approach for slight perturbation in control variables. Such perturbation in control variables aims towards the improvement of response.

5.1.4 Influence of Knowledge of Statistics

The calculations are simple in an EVOP analysis. The analysis requires elementary statistics, DOT diagram,[5] random variation, frequency distribution, distribution characteristics (mean and variance), standard deviation, and standard errors.

5.2 Classification of EVOP

Important classes of EVOP are discussed in subsequent paragraphs.

5.2.1 *Classical EVOP*

This is the first suggestion by Box and Draper.[6] Classical EVOP is discussed with examples in this book. Planned factorial design points are the basis of this method and consider the operation on scientific feedback.

5.2.2 Rotating *Square EVOP*

Harrington[7] suggested rotating square EVOP. The ROVOP method uses the concept of 2^2 EVOP as the basis. If the experiment does not give improvement in the process, the square in initial conditions rotates at 45°. If this fails, a third square assumes similarly within *second square* by following the earlier process (Figure 5.3). If significant improvement appears, the design progresses cautiously in the best direction.[6]

5.2.3 *Random EVOP*

This method is proposed by Satterthwaite.[8] Experimental conditions do not use planned factorial design condition. This is unlike for classical EVOP.

5.2.4 Simplex *EVOP*

Spendley et al.[9] suggested the experimental process using the movement of a regular simplex (Figure 5.4). For a two-variable case, an equilateral triangle is considered and experiments are conducted based on conditions at three

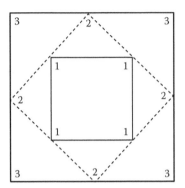

FIGURE 5.3
Rotating square EVOP design plan.

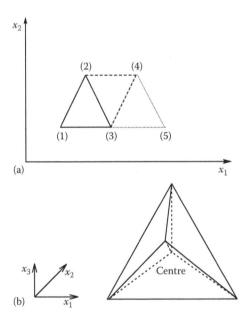

FIGURE 5.4
Simplex EVOP design plan: (a) in two dimensions and (b) in three dimensions.

vertices. If the experimental result shows that vertex 2 conditions give the worst result, then a vertex opposite to vertex 2 is added. This forms a new equilateral triangle. Thus, experiments continue further. If the number of variables is k, regular geometric figure of $(k + 1)$ is the simplex. This gives empirical feedback.

5.3 Specific Terminologies

Cycle and phase are important terms in the EVOP method.

Cycle: A single yield/response of a complete operating condition is made in a cycle. For example, chitinase production in a batch reactor is influenced by aeration rate, pH (controlled), and agitation.

Phase: Repetition of cycle of operation conditions is defined as a phase in the EVOP.

5.4 Worksheet for EVOP

The worksheet is based on a concept proposed by Box and Draper.[6] There are several parts in the calculation. They are described chronologically in this section.

Part 1: Geometric presentation of the first set of experiments is described in Figure 5.5.

Corners indicate the response at a suitable combination of variables 1 and 2, that is, the operating conditions.

Let us assume the response for cycle number 1.

Part 2: Calculations

A. Calculation of averages

Table 5.1 gives the detailed calculation pertaining to this method.

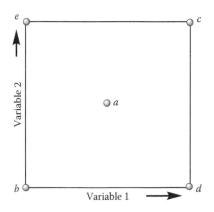

FIGURE 5.5
Geometric presentation of the first set of experiments.

TABLE 5.1

Summary of Calculation

Items	Description	Operating Conditions				
		a	b	c	d	e
1.	Sum for previous set of run	Y_1 (or NI)	Y_2 (or NI)	Y_3 (or NI)	Y_4 (or NI)	Y_5 (or NI)
2.	Average of previous set of run	Y_1 (or NI)	Y_2 (or NI)	Y_3 (or NI)	Y_4 (or NI)	Y_5 (or NI)
3.	Present observation (response)	Y_{a1}	Y_{b1}	Y_{c1}	Y_{d1}	Y_{e1}
4.	[a]Difference of values in item (2) – values in item (3)	$(Y_1 - Y_{a1})$	$(Y_2 - Y_{b1})$	$(Y_3 - Y_{c1})$	$(Y_4 - Y_{d1})$	$(Y_5 - Y_{e1})$
5.	New total of item (1) and item (3)	$(Y_1 + Y_{a1})$ or Y_{a1}	$(Y_2 + Y_{b1})$ or Y_{b1}	$(Y_3 + Y_{c1})$ or Y_{c1}	$(Y_4 + Y_{d1})$ or Y_{d1}	$(Y_5 + Y_{e1})$ or Y_{e1}
6.	New average of response Y_{avi}, values of item (5)/number of values (i.e. $(Y_1 + Y_{a1})/2$ or Y_{av})	$Y_{av\,a}$	$Y_{av\,b}$	$Y_{av\,c}$	$Y_{av\,d}$	$Y_{av\,e}$

[a] Highest positive and negative differences are highlighted, which gives the range stated in Section C.

NI: If the result is not available/no information for cycle 1, the space is blank.

B. *Calculation of effects*

Effect of variable $1 = 1/2\,(Y_{avd} + Y_{avc} - Y_{avb} - Y_{ave})$

Effect of variable $2 = 1/2\,(Y_{avc} + Y_{ave} - Y_{avb} - Y_{avd})$

Combined effect of variables 1 and $2 = 1/2\,(Y_{avb} + Y_{avc} - Y_{avd} - Y_{ave})$

Change in mean effect $= 1/5\,(Y_{avb} + Y_{avc} + Y_{avd} + Y_{ave} - 4Y_{ava})$

C. *Standard deviation for*

Prior estimate of standard deviation = 1.8.

Sub-step 1: Sum for previous set of run

Sub-step 2: Average of previous set of run

For item 4 of Table 5.1, highest positive and negative values give the range (=R)

New sum = Range $\times f_{k,n}$, where $f_{k,n}$ is the constant obtained from a chart (Table 5.2 and Appendix A.7)

New average = New sum/$(n_c - 1)$

D. *2 × error limits for*

New average $= (2/\sqrt{n_c}) \times \sigma$

New effect $= (2/\sqrt{n_c}) \times \sigma$

Change in mean $= (1.78/\sqrt{n_c}) \times \sigma$

TABLE 5.2

Data for $f_{k,n}$

Number of Cycles = n	k = Number of Sets of Conditions in Block 2 ... 10
2 to 20	Values are available in Box and Hunter.[2]

Source: Box GEP and Hunter JS, *Techometrics*, 1, 77–95, 1959.

where:

n_c is the number of cycle

E. The justified values for averages and effects should satisfy 2 × standard error.

Part 3: Suggestion of information board

Following suggestions can be proposed for the process:

• The process may continue without change in process conditions.

or

• The process may require modification of process conditions. The options are as follows:

  – Change in one of the conditions

  – Continue in the projected favourable direction

  – Substitution of old conditions

  – Modification of type of variables

Part 4: Application of Yates' algorithm

For the calculation of effects of the variables, Yates' algorithm[10] gives added reference conditions. The sum of squares check is used in the calculation. If the calculation shows four times the value, then it justifies the correct composition.

5.4.1 Solved Examples

Worked out examples explain the theory.

Example 5.1

Enzyme production at five different conditions is given in Figure 5.6 and Table 5.3.

Calculation:

Effect of pH = $1/2 \, (y_3 + y_5 - y_2 - y_4) = 0.0131$
Temperature effect $= 1/2 \, (y_4 + y_3 - y_2 - y_5) = -0.077$
Combined effect of temperature and pH = $1/2 \, (y_2 + y_3 - y_4 - y_5) = 0.001$
Change in mean effect $= 1/5 \, (y_2 + y_3 + y_4 + y_5 - 4y_1) = -0.06972$
pH effect = $1/2 \, (y_3 + y_5 - y_2 - y_4) = 0.017$

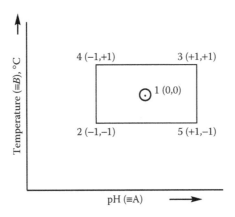

FIGURE 5.6
Enzyme production.

TABLE 5.3

Coded and Decoded Values of Temperature and pH

Conditions (temperature, °C; pH)	1 ($\equiv 0, 0$)	2 ($\equiv -1, -1$)	3($\equiv 1, 1$)	4($\equiv -1, 1$)	5($\equiv 1, -1$)
Observed activity (y_i), U	0.3914	0.3367	0.2728	0.2587	0.3488

TABLE 5.4

Summary of Average and Sums Calculation

Description	1	2	3	4	5
Previous cycle sum	0.3914	0.3367	0.2728	0.2587	0.3488
Previous cycle average	0.3914	0.3367	0.2728	0.2587	0.3488
New observations	0.3925	0.3370	0.2823	0.2589	0.3540
Differences	−0.0011	−0.0003	−0.0095	−0.0002	−0.0052
New sums	0.7839	0.6737	0.5551	0.5176	0.7028
Averages	0.3920	0.3369	0.2776	0.2588	0.3514

Temperature effect $= 1/2\ (y_4 + y_3 - y_2 - y_5) = -0.076$
Combined effect of temperature and pH $= 1/2\ (y_3 + y_2 - y_4 - y_5) = 0.0022$
Change in mean effect $= 1/5\ (y_2 + y_3 + y_4 + y_5 - 4y_1) = -0.07$
Range $= -0.0002 + 0.0095 = 0.0093$ (refer Table 5.4)

New $S = $ Range $\times f_{k,n} = 0.0028$
    Therefore, average $= $ New sum/$(n-1) = 0.0028$

Error limits:

    For new average $= 0.004$
    For new effects $= 0.004$
    For change in mean $= 0.0035$

Example 5.2

Analysis of Example 5.1 with two sets of runs is necessary by the EVOP procedure. Table 5.5 gives operating conditions with the corresponding responses.

Table 5.6 gives coded values of temperature and pH for β-glucanase production.

Solution:

Figure 5.6 shows the EVOP plan.

Calculation of effects:

For A, that is, pH:

i. The change of a variable from its lower value to its higher value (2 → 5), in the presence of lower value of B

Effect $= (Y_5 - Y_2)$

ii. The change of a variable from its lower value to its higher value (4 → 3), in the presence of higher value of B

Effect $= (Y_3 - Y_4)$

Therefore, the effect of A $= 1/2\ (Y_5 + Y_3 - Y_2 - Y_4)$

For B, that is, temperature:

i. Similar change for a variable from its lower value to higher value (2 → 4), in the presence of low value of A

Effect $= (Y_4 - Y_2)$

TABLE 5.5

Two Sets of Responses for β-Glucanase Production

	1	2	3	4	5
Operating Conditions	(0, 0)	(−1, −1)	(+1, +1)	(−1, +1)	(1, −1)
β-glucanase produced	0.3914	0.3367	0.2728	0.2587	0.3488
New observations for β-glucanase produced	0.3925	0.3370	0.2823	0.2589	0.3540

TABLE 5.6

Coded and Decoded Values of Temperature and pH for β-Glucanase Production

	−1	0	+1
Temperature	25°C	30°C	35°C
pH	4	5	6

TABLE 5.7

Summary of Data for Cycle 1

Operating Condition	1	2	3	4	5
(i) Previous sum for block experiments	Nil	Nil	Nil	Nil	Nil
(ii) Previous average for block experiments	Nil	Nil	Nil	Nil	Nil
(iii) New condition for block	0.3914	0.3367	0.2728	0.2587	0.3488
(iv) Difference of (ii) and (iii)	**0.3914**	0.3367	0.2728	**0.2587**	0.3488
(v) New sum for block	0.3914	0.3367	0.2728	0.2587	0.3488
(vi) New average for block	0.3914	0.3367	0.2728	0.2587	0.3488

Note: Boldfaced values are the highest and lowest values among five conditions. This is as per item (4) of Table 5.1. Those values are necessary to calculate the range.

    ii. Similar change for a variable from its lower value to higher value (5 → 3), in the presence of higher value of A

$$\text{Effect} = (Y_3 - Y_5)$$

Therefore, effect of $B = 1/2\,(Y_3 + Y_4 - Y_5 - Y_2)$

For AB:

    i. The effect at lower level of A and B

$$AB = (Y_2 - Y_5)$$

    ii. The effect at higher level of A and B

$$AB = (Y_3 - Y_4)$$

Therefore, the mean effect of $AB = 1/2\,(Y_2 + Y_3 - Y_4 - Y_5)$

Standard deviation for cycle 1 = 0.04935120

Calculation of effects for cycle 1. Data are given in Table 5.7.

$$\text{Effect}(A) = \frac{1}{2}\left(Y_5 + Y_3 - Y_2 - Y_4\right) = \frac{1}{2}\left(0.6216 - 0.5954\right)$$

$$E(A) = 0.0131$$

$$\text{Effect}(B) = \frac{1}{2}\left(Y_3 + Y_4 - Y_5 - Y_2\right) = \frac{1}{2}\left(0.5315 - 0.6855\right)$$

$$E(B) = -0.077$$

$$\text{Interaction effect }(AB) = \frac{1}{2}\left(0.6095 - 0.6075\right) = 0.001$$

$$\text{Change in mean effect} = \frac{1}{5}\left(Y_2 + Y_3 + Y_4 + Y_5 - 4Y_5\right)$$

$$= \frac{1}{5}\left(1.217 - 1.5656\right) = -0.06972$$

TABLE 5.8

Results for Cycle 2

Operating Condition	1	2	3	4	5
(i) Previous sum for block	0.3914	0.3367	0.2728	0.2587	0.3488
(ii) Previous average for block	0.3914	0.3367	0.2728	0.2587	0.3488
(iii) New condition for block	0.3925	0.3370	0.2823	0.2589	0.3540
(iv) Difference of (ii) and (iii)	−0.0011	−0.0003	**−0.0095**	**−0.0002**	−0.0052
(v) New sum for block	0.7839	0.6737	0.5551	0.5176	0.7028
(vi) New average for block	0.39195	0.33685	0.27755	0.2588	0.3514

Note: Boldfaced values in this table are the highest and lowest values among five conditions. This is as per item (4) of Table 5.1. Those values are required to calculate the range.

$$\text{Range of cycle I for block as per data in Table 5.7} = (0.3914 - 0.2587)$$
$$= 0.1327$$

Standard deviation for cycle 2 = 0.048865831
Calculation of effects for cycle 2. Data are given in Table 5.8.

$$\text{Effect } (A) = \frac{1}{2}(Y_5 + Y_3 - Y_2 - Y_4) = \frac{1}{2}(0.62895 - 0.59565)$$
$$E(A) = 0.01665$$

$$\text{Effect } (B) = \frac{1}{2}(Y_3 + Y_4 - Y_5 - Y_2) = \frac{1}{2}(0.53635 - 0.68825)$$
$$E(B) = -0.07595$$

$$\text{Effect } (AB) = \frac{1}{2}(Y_2 + Y_3 - Y_4 - Y_5) = \frac{1}{2}(0.6144 - 0.6102) = 0.0021$$

$$\text{Change in mean effect} = \frac{1}{5}(Y_2 + Y_3 + Y_4 + Y_5 - 4Y_5) = \frac{1}{5}(1.2246 - 1.5678) = -0.06864$$

$$\text{Range for cycle 2} = (-0.0002 + 0.0095)$$
$$= 0.0093$$

$$f_{k1n} = f(5,2) = 0.30$$

Therefore

New sum S_1 = range $\times f_{k,n}$ = 0.0093 × 0.30 = 0.00279
New average = $(S_1/2n-3)$ = (0.00279/4−3) = (0.00279)
New sum of all block = 0.00279
New average of all block = 0.00279 = σ

Calculation of error limits

For average = $\pm 2\,\sigma/\sqrt{n}$ = ± 2 × (0.00279/√2) = ± 2 × 0.00394565
For effects = $(\pm 2\,\sigma/\sqrt{n}) \times 0.71$ = ±0.00280142
For change in mean = $(\pm 2\,\sigma/\sqrt{n}) \times 0.63$ = ±0.00248576

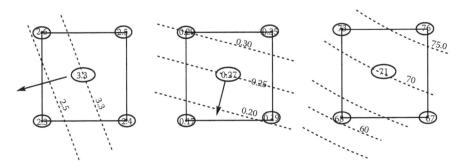

FIGURE 5.7
Response surfaces achieved in the EVOP plan.

5.5 Response Surface

From the EVOP experimental strategy, it is possible to draw response surfaces. A few examples are shown in Figure 5.7. Box and Draper[6] described a detailed analysis of responses achieved in the EVOP experimental plan.

Exercises

5.1 Experimental conditions are given in the Figure 5.1.1.

1, 2, 3, 4, and 5 represent the conditions of the experimental runs. Table 5.1.1 summarizes the responses in the two sets of experiments.

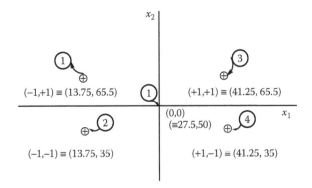

FIGURE 5.1.1
Two variables x_1 and x_2 in chitinase biosynthesis. Coded and decoded values of the variables are given in the figure.

TABLE 5.1.1

Responses of Chitinase in Two Cycles

Conditions	1	2	3	4	5
Cycle 1	0.54	0.45	0.43	0.47	0.43
Cycle 2	0.59	0.48	0.53	0.39	0.40

Calculate the individual effects of x_1 and x_2, interaction effect of x_1 and x_2, and mean effect. Suggest the EVOP information after cycle 2. All relevant standard deviation calculations are necessary.

5.2 Chitinase was produced under controlled pH, aeration, and agitation conditions. Table 5.2.1 summarizes production conditions and responses.

Calculate the effects of A, B, C, AB, BC, CA, and ABC for each block. Follow the EVOP procedure. Recommend the production strategy.

5.3 Table 5.3.1 shows the sets of conditions of a 2^3 factorial design for a biopolymer synthesis in a bacterial strain.

Use these data for EVOP optimization; Figure 5.3.1 is relevant. Estimate the main effects, two-factor, and three-factor interactions. Calculate variance and standard error. Suggest the decision of the design experiments.

TABLE 5.2.1

Chitinase Production Following the EVOP Strategy

Run Number	Block Number	pH Controlled ($\equiv A$)	Aeration Rate $\left(\dfrac{m^3\,\text{Air}}{m^3\text{Medium (min)}} \right)$ ($\equiv B$)	Agitation Rate (rev/min) ($\equiv C$)	Response (Chitinase, U)
	1				
1		4	3.67	350	1.22
2		8	1	150	0.89
3		8	3	300	1.03
4		3.3	6.47	364	1.19
5		2.2	5.08	188	1.19
	2				
6		6.3	5.45	234	1.28
7		5	1	300	0.98
8		3	6.5	400	1.04
9		8	1	300	1.00
10		5	3	150	1.15

TABLE 5.3.1

Biopolymer Synthesis in a Bacterial Strain under Different
Fermentation Conditions

Standard Order	Soy Extract (kg/m³)	Peptone (kg/m³)	NaCl (kg/m³)	Biopolymer (kg/m³) × 10³
1	4	4	3	5.93
2	23	4	3	6.94
3	4	23	3	2.72
4	23	23	3	5.41
5	4	4	17	6.8
6	23	4	17	8.91
7	4	23	17	5.29
8	23	23	17	4.54

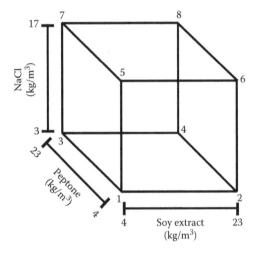

FIGURE 5.3.1
Data for three variables to be optimized using the EVOP design technique.

5.4 Suppose in a fermentation experiment temperature (A), pH (B), and phenylacetic acid (C) influence the synthesis of penicillin amidase using *Escherichia coli*. Experimental runs are in two blocks, which represent the geometric figure with a common geometric centre as in Figure 5.4.1. The range of A, B, and C are 20°C to 35°C, 4 to 7, and 0.1 to 0.25 kg/m³.

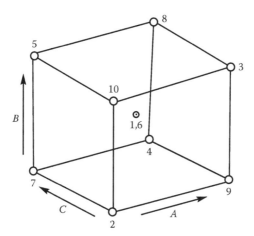

FIGURE 5.4.1
Experimental runs for three variables (*A*, *B*, and *C*). Experimental conditions are vertex of the geometric figure.

5.5 Extraction of penicillin amidase from *E. coli* cells using chloroform depends on temperature and the concentration of chloroform. Four experimental conditions are represented in the vertices of the block shown in temperature versus concentration. Three separate EVOP strategies are (a), (b), and (c) in the Figure 5.5.1. Conclude the option with justification for the EVOP recommendation.

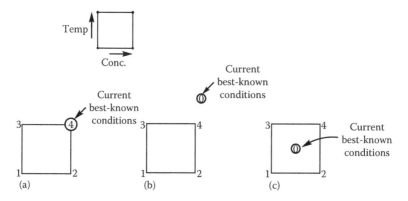

FIGURE 5.5.1
Extraction of penicillin amidase strategies of EVOP. (a) Best condition is one of the vertex of the block; (b) best condition is outside the block; (c) best condition is the centre of the box.

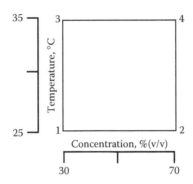

FIGURE 5.6.1
Conditions of biopolymer extracted from cell.

5.6 Purity of a biopolymer extracted from a cell depends on the concentration of hypochlorite (30% v/v to 70% v/v) and temperature between 25°C and 35°C. The conditions need to be optimized by the EVOP process. Figure 5.6.1 gives the conditions of experiments.

5.7 An intracellular enzyme synthesis is influenced by temperature between 25°C and 35°C and phenyl acetic acid between 0.2 and 0.25 (kg/m³). The range of enzyme synthesis is between 0.02 and 0.5 U/g dry weight of cell as per Figure 5.7.1. Use EVOP to suggest the future plan.

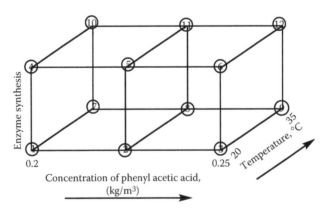

FIGURE 5.7.1
Intracellular enzyme synthesis.

References

1. Box GEP, Some theorems on quadratic forms applied in the study of analysis of variance problems, II, Effects of inequality of variance and of correlation between errors in the two-way classifications, *Annals of Mathematical Statistics*, **25**, 16–60, 1954.

2. Box GEP and Hunter JS, Condensed calculations of evolutionary operation programs, *Techometrics*, **1**, 1, 77–95, 1959.

3. Banerjee R and Bhattacharyya BC, Evolutionary operation (EVOP) to optimize three dimensional biological experiments, *Biotechnology and Bioengineering*, **41**, 67–71, 1993.

4. Bruns RE, Scarminio IS, and de Barros Neto B (Eds.), *Statistical Design – Chemometrics*, Elsevier, Amsterdam, the Netherlands, 2006.

5. Montgomery DC (Ed.), *Design and Analysis of Experiments*, 7th edition, John Wiley & Sons, Delhi, India, 2009.

6. Box GEP and Draper NR (Eds.), *Evolutionary Operation*, John Wiley & Sons, New York, 1969.

7. Lowe CW, Some techniques of evolutionary operations. *Transactions of the Institutions of Chemical Engineers*, **42**, T332–T344, 1964.

8. Satterthwaite FE, Reply to discussion of papers FE Satterthwaite and TA Budne on "Random Balance", *Techometrics*, **1**, 185, 1959.

9. Spendley W, Hext GR, and Himsworth FR, Sequential application of simplex designs in optimization and EVOP, *Techometrics*, **4**, 441–461, 1962.

10. Johnson NL and Leone FC (Eds.), *Statistics and Experimental Design in Engineering and Physical Sciences*, Vol II, John Wiley & Sons, New York, 1977.

6

Taguchi's Design

OBJECTIVE: This chapter intends to make the reader familiar with the basics of the Taguchi design of experimentation, which is very popular in the field of biological systems. The reader can better understand the basic principles based on which the Taguchi design works, along with its advantages and the disadvantages compared to other experimental and non-experimental statistical techniques popularly used to optimize biological systems.

In addition, the reader may be able to design a Taguchi design of experiment to optimize any appropriate system and interpret the results of a system, which has been analyzed using the Taguchi design. The intention is to give the reader a good interest to look into this design option when needed in practice.

6.1 Introduction

Among the multiple statistical approaches that are available in order to optimize the systems, the concept of design of experiments is a popular and a widely used technique, mainly intended for systems that cannot be modelled mathematically based on theoretical concepts. These systems, generally, require the observer to conduct a set of experiments, which lets the observer model the system.

However, there are multiple ways to design the experiments, which give us the insight towards the nature of the system, which is being observed. One among the many available approaches is the Taguchi design, which is used extensively[1] – more so in the case of biological systems.

The Taguchi design is one of the most robust designs of experimentation that is available. It is a very well-studied design that has very clear guidelines of which experimental points to pick and which ones to avoid in order to get a reasonable amount of information from the system that is being analyzed.[1] It is one of the very few experimental systems that is laid out and suitable from an industrial perspective, considering the deviations from ideality. Most other design only look at systems from an experimental perspective, leaving the accuracy of the final results many a times at the mercy of the deviations that occur from ideality due to practical causes mostly out of our control.[2]

It is very structured and analyzes the parameters that can be optimized in terms of the nature of the disturbances, arising from a given set point either external, planned, or unplanned (inaccuracies in results for repeat experiments).[3]

It is necessary that a minimum number of well-balanced experiments are performed with reduced variance having optimal settings of control parameters to get the best result of desired output (the response). Taguchi's technique is based[2] on such a concept. Dr. Taguchi suggested a method based on *orthogonal array* and experiments for multi-variable system in which the objective functions for optimization is *signal-to-noise* (S/N) ratio.[2,3] The S/N ratio is defined as the logarithmic value of the desired output of the experiment. Taguchi design of experiments can be categorized as four sub-design components. They are robust design, concept design, parameters design, and tolerance design.[2,3]

6.1.1 History of Taguchi's Design

Dr. Taguchi developed methods to improve the quality with which products are manufactured.[1] Many organizations have shown interest in Taguchi's work.

Taguchi worked to show that using the concept of design of experiments, it is possible not only to improve the quality of the process, but it is also economically profitable. This method appears to be user-friendly. It is possible to quantify every parameter involved in the process. Taguchi's quality loss function and the concept of robust design are of great use.[1]

Before we look at how different Taguchi's perception of quality was, we should get a brief idea of how any industry that produces a product perceived quality at that time. Any industry has systems that monitor the quality of a final outcome reaching to the customer. From Taguchi's concept, a quality loss is defined as in Figure 6.1.

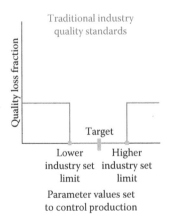
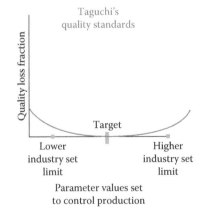

FIGURE 6.1
Taguchi's quality concept.

As per this concept, quality[4] is the ability of the product to perform the design function throughout its life under reasonable operating conditions.

6.2 Aim of Taguchi's Design

For a *robust* process, the planned minimal experiments have minimal variation in the output (the response) in the presence of noise (the disturbance/ the perturbation). The design is insensitive to component variation in environmental conditions and a minimal variation about the target value (the response).[2]

6.3 Experimental Designs versus Taguchi's Design

It is well known that when modelling systems, the use of experiments is very valuable and is unavoidable in some cases. Therefore, to conduct the experiments with various factors, varying at different levels is rather important. The dependence of the response of the system with each of the factor guides the experimental plan to obtain results in confidence.

Let the various levels of the factors be represented by 1, 2, 3, and so on by simple numeric. The factors that are being considered are represented by alphabets, namely, A, B, C, and so on. Therefore, a representation of A1 means that it refers to factor A being set at level 1.

Ideally, every system is being controlled only by a single factor; that is, one can simply change that factor and control the system. In addition, the dependence of the response with respect to that factor will not be very difficult to understand. Unfortunately, systems that exist seldom are of the stated example. Most systems have multiple factors affecting their performance – the number of factors that affect the system could be more than 10 in some cases. For the sake of discussion, cases of a few variables are considered, where each variable can be varied at three different levels, for example, 1, 2, and 3. Such systems are very common, especially among biological systems.

The factorial design of experiments discussed before is one of the most extensive methods used for designing experiments. It suffers from a very important issue of being over extensive and leading to multiple experimentation.[4] Therefore, an evolved form of experimentation introduced and used widely is the fractional factorial design, which as the name suggests, uses only a part of the factorial design and models the system. It is always a tricky situation to decide which of the levels of the factors are to be used for multiple levels. In most of the situations, the fractional factorial design is unbalanced.[4]

Taguchi's design is unique in being able to look at the variables or parameters of experimentation from both the perspectives of the set point and the disturbance. Taguchi's method uses the concept of orthogonal arrays.[1]

6.4 Basis of Taguchi's Design Technique

In general, there are three major basis of this technique[1]:

- Highly fractionated factorial design
- Orthogonal array
- Application of some statistical methods to solve problems

6.5 Classes of Optimization Problems

In Taguchi's method, the problem is either static or dynamic.[2]

1. *Static problems*: For a bioprocess influenced by a number of controlled variables, the optimization involves defining best complete factors to achieve desired output, irrespective of the presence of the noise, assuming that the noise does not influence the output (response) (*cf.* Figure 6.2).

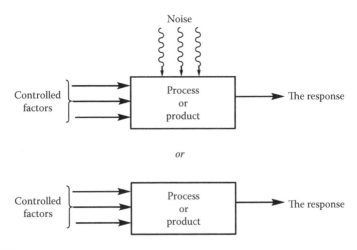

FIGURE 6.2
Schematic of static problem.

2. *Dynamic problem*: If the system for optimization has a signal, that influences the output along with the controlled factors; the optimization involves both signal input and controlled factors to get the *best* output (the response), provided the desired ratio of input signal to output is nearest to the desired values.[2]

6.6 Terminologies

1. *Parameters*: Taguchi calls factors[2] that can be controlled during the process as parameters. For a biological process, temperature, pH, agitation, and aeration are controlled, which are parameters.
2. *Noise*[2]: Factors that are difficult to control, but influence the output are called noise, for example, variation in quality of raw materials, organism's physiology, and morphology.

6.7 Array in Orthogonal Design

In Taguchi's design, the arrangement of parameters and noise is in two different layers: inner and outer arrays.[1]

Inner array: This involves only the parameters.

Outer array: This involves the noise source.

What exactly is an orthogonal array?

In mathematical terms, the word *orthogonal* means perpendicular. In a more crude way, one can interpret it as being independent. The underlying idea of using orthogonal arrays is that different arrays are independent. If these are mapped onto the variables that control the production or the output of a system, then clearly the experimental points are independent,[4] which will indirectly imply that we do not waste our experimental resources by doing experiments that give the same information like that was given by another experiment. To emphasize the above point, consider the following orthogonal arrays.

1 1 1 1 1 1 and 1 1 1 2 2 2 are arrays that are orthogonal to each other. Here we have considered a system whose controlling parameters can be represented by an array with 6 bits. The parameters are considered to be varied at two different levels: 1 and 2. By this, we can see that if we were to plot it on a seven-dimensional space, then these two vectors would be perpendicular

to each other. In that space, the origin would be 1.5, 1.5, 1.5, 1.5, 1.5, 1.5, thus proving that the two vectors are in unrelated directions and cannot be resolved into components that can point in the same direction, making us know the most about this space.

In other way, assume that we are shown a six-dimensional space with a given reference point (origin) and are told that we could select two vectors from the origin with the objective of discovering or defining the seven-dimensional spaces to the best extent possible. In such a situation, clearly our choice would be to use two perpendicular vectors. This helps us to avoid getting repetitive information while extracting the maximum information possible.

Mapping the above analogy on the experimented system, we considered that is like the seven-dimensional space with the different parameters having different dimensions. Therefore, an orthogonal experimentation design will use the minimum resources[1] available to get more information out of the system for limited number of experiments.

The most important constraint is that the different experimental points are independent of each other. If there is repetition of information obtained from the experiment, the experimental design is no longer orthogonal.[1]

Example 6.1

Consider a system with say seven variables and two levels. For this system, we need to use eight experiments to analyze the system. How do we represent this design?

Solution:

The systematic solution is as follows.

- The base is 2 and the exponent is 7. Therefore, the first part is (2^7).
- The term L is used to represent Latin square.
- The number of experiments to be done is 8 so, $n = 8$.
- Thus the final representation is $(2^7) L$–8.

It is very important to be able to clearly represent such systems in a universal method.[1] This leads to lesser confusion. Also, it lets the user/experimenter to be clear about what the experimenter has been doing in case of confusion. This is especially important when the system is very big in terms of number of levels and factors, which is quiet often the case in realistic biological systems. This universal way of representing also leads to clear mentioning of many properties of the system under Taguchi's design.[1]

However, it is important to know the advantages of universal representation of orthogonal array. It leads to lesser confusion among various scientists. Consider the case where you come up with a design of experiment for a particular system; in most industrial scenarios, you will not be the person who is directly interested in a particular system, but it will be the client who comes to you (who is an expert in design of experiments). The client will have very little know-how to understand and to explain what has been done to other experts in the future for other developmental studies.

The array tells us the number of trials required for modelling the system. This way of representing a system according to universal convention only obviously helps people who do not have much knowledge about the system.

The other details it provides are the number of factors that are controlled by us and the actual design that is being used – the above representation indicates directly the use of Taguchi's design using orthogonal arrays.

Advantages of Taguchi's Design[1]

1. It is very simple to use. Since it has been used for many decades, all information regarding the design is also easily available and present in abundance.
2. This is one of the most reliable designs, with very clearly documented information being available.
3. It is one of the most robust designs. One can get the most information to be worked upon for any given design of experimentation for a limited number of experiments. For a system with eight variables, having each variable being able to take up three different levels, the number of factorial experiments needed would be 3^8, which is equal to 6561 experiments. On the other hand, if one looks at the same system with the help of Taguchi's design of experiments, the number of experiments needed would be 18! Clearly, the Taguchi's design saves a lot of time and resources with more possibility of predictions regarding the system compared to a factorial design of experiments.

Taguchi's concept of quality loss function helps us achieve a target quality.

How exactly does one use these orthogonal arrays for one's experiments?[6]

Orthogonal array can lead to simpler experiments and easier calculations. One particular array may be suitable for a number of experimental conditions. There are online calculators that can directly give the experimental grid. The most commonly used experimental grids for Taguchi designs have been mentioned in Section 6.7.1, but since it is not possible to cover all the possible designs, the reader is advised to search online tables for the system.

What do I do when the system does not have the same number of variables that I find in the grid? Does this mean that the system cannot be solved by Taguchi's design?[2]

Lets assume that the system contains five variables. In this case, $L-4$ orthogonal array cannot be used for this purpose, as it is a design meant for only three variables. What should one do then? The next array is the $L-8$ array. This array gives one the chance to work on a system with seven variables, but there are only four variables. The answer is to make space for three dummy variables. When the turn comes for those experiments to be performed, it is better to repeat the experiment, if there is a question of reliability or random errors. Otherwise, the response is rewritten in the next row. To perform analysis automatically, the effect of the dummy variable will come to be zero; if the response is just re-recorded, but if all the dummy variable effects are coupled, probably there will be big error contribution. Taguchi's design can be extended to any number of variables defining the system for experiment.

6.7.1 Sample Design Arrays for Taguchi's Experimental Plan

The most commonly used Taguchi's orthogonal arrays (Tables 6.1 through 6.3) are given with a view to implement Taguchi's experimental plan (see Figure 6.4). It is easy to cross-check if the given arrays are orthogonal. Consider a dot product of both the different arrays, if it is zero, which will help one to conclude that the two vectors are perpendicular indeed. It has to be noted that the

TABLE 6.1

L-4 Orthogonal Array

Experimental Run Numbers	pH	Temperature (°C)	Agitation Rate (rev/min)
1	1	1	1
2	1	2	2
3	2	1	2
4	2	2	1

TABLE 6.2

L-8 Orthogonal Array

Experimental Run Numbers	Glucose	Peptone	$CaCl_2$	$FeSO_4$	NH_4NO_3	Tween 80	Vitamin B_{12}
1	1	1	1	1	1	1	1
2	1	1	1	2	2	2	2
3	1	2	2	1	1	2	2
4	1	2	2	2	2	1	1
5	2	1	2	1	2	1	2
6	2	1	2	2	1	2	1
7	2	2	1	1	2	2	1
8	2	2	1	2	1	1	2

TABLE 6.3

L-9 Orthogonal Array

Experimental Run Numbers	Chitin	Malt Extract	KH_2PO_4	$ZnCl_2$
1	1	1	1	1
2	1	2	2	2
3	1	3	3	3
4	2	1	2	3
5	2	2	3	1
6	2	3	1	2
7	3	1	3	2
8	3	2	1	1
9	3	3	2	3

vectors joining the adjacent different experimental points are the perpendicular ones and not the vectors joining origin to these points. Note also that the origin here is an array constituted by the mid-point of the experimental levels; for example, if they have three levels, namely 1, 2, 3, origin would be 2,2. Therefore, this has to be adjusted for a while taking the dot product.

6.8 Signal-to-Noise Ratio

This ratio is the change in response due to the change in the parameter or signal value to the change in the response due to unknown noise effects.[2,4]

Consider that a system operates at a particular experimental point, having S/N ratio of A. Suppose that the experimental point is moved to another point of experiment in the hope of improving the target production with the S/N ratio of B.

What does it really mean?

To answer that question one has to keep in mind the unique Taguchi's quality definition. If say, the change in set point leads to higher production target, say if the S/N ratio of A is more than B, the environmental noises will lead to a bigger or more oscillations from the set target in case of B. According to Taguchi, this means more loss in quality even though the actual target is higher. In this context, it is necessary to interpret and to use the concepts of S/N values.

The S/N ratio tells how reliably the production happens at the given target point. It does not depend on the actual target point. For example, consider the production targets for 20 units with deviations of +4 or −4, and, production targets around 40 units with deviations of +6 or −6. The first choice is considered better even though the set target is much lower. This is Taguchi's quality definition and the S/N ratio gives a measure of it. It does not exactly quantify this, but it takes into account how the target varies when the parameter varies with a change in the noise level.

To determine the S/N ratio, a different set up of experimental design is considered to introduces the concept of inner and outer arrays, which are additional components in the Taguchi design.

The inner arrays include the conventional Taguchi's design arrays for the process variables that lead to change in the responses, that is, the signal parameters. The outer array includes the effect caused by the noise factors. They are not designed orthogonally, but are factorial in nature, which makes it possible to not to miss the interaction effects in the case of the inner array elements.

For every run, the analysis of S/N ratio is to select the best run conditions. The recommendation of Taguchi is to maximize the relation of S/N value[2–7] (Equation 6.1).

$$\frac{S}{N} = 10 \ \log_{10} \frac{\left(\text{Mean response for a run}\right)^2}{\text{Variance}} \tag{6.1}$$

For static and dynamic problems, the S/N ratio is defined separately for optimization studies. However, Taguchi has proposed a large number of S/N ratios.

For static problems, the three ideologies of describing S/N ratio are as follows:

- Smallest possible S/N value is better, that is, ideally value is zero.
- S/N ratio is calculated based on the reciprocal value used in *smaller-the-better*.
- Nominal value of S/N ratio is the best, that is, the expression is as per Equation 6.1.

In dynamic problems, S/N ratio is based either on slope or on linearity of input-to-output ratio. In the Equation 6.1, S/N ratio contain a term in the numerator which is the square of the slope of input-to-ouput ratio.

Hypothetical Example 6.1

Consider an *L*-8 array (Table 6.4).

This matches the outer array requirements but the inner array is different. Assuming there are two different noise factors, making up the parameters in the inner array. Also, let us assume that each of those can take up two different levels. In such a case, four different combinations are possible with these disturbance levels. Therefore, we create each of these combinations artificially and then perform the outer array of experiments for all these combinations. In this case, the number of experiments is $= 4 \times 8 = 32$ experiments. Table 6.5 is the response table.

Modified form of Equation 6.1 is

- When we are optimizing the response characteristic,[3]$SN_i = 10 \times \log_e(y_i'^2/s_i^2)$
- When we are minimizing the response characteristic, $SN_i = -10 \times \log_e[\Sigma_j (y_{i,j}^2/N_i)]$
- When we are optimizing the response characteristic, $SN_i = 10 \times [1/N_i] \times \log_e[\Sigma_j (1/y_{i,j}^2)]$

TABLE 6.4

L-8 Orthogonal Array

Experiment	Glucose	Peptone	$CaCl_2$	$FeSO_4$	NH_4NO_3	Tween 80	Vitamin B_{12}
1	1	1	1	1	1	1	1
2	1	1	1	2	2	2	2
3	1	2	2	1	1	2	2
4	1	2	2	2	2	1	1
5	2	1	2	1	2	1	2
6	2	1	2	2	1	2	1
7	2	2	1	1	2	2	1
8	2	2	1	2	1	1	2

TABLE 6.5

Response for Taguchi's Experiment

Experimental Run Numbers	R_1	R_2	R_3	R_4
1	Y_{11}	Y_{12}	Y_{13}	Y_{14}
2	Y_{21}	Y_{22}	Y_{23}	Y_{24}
3	Y_{31}	Y_{32}	Y_{33}	Y_{34}
4	Y_{41}	Y_{42}	Y_{43}	Y_{44}
5	Y_{51}	Y_{52}	Y_{53}	Y_{54}
6	Y_{61}	Y_{62}	Y_{63}	Y_{64}
7	Y_{71}	Y_{72}	Y_{73}	Y_{74}
8	Y_{81}	Y_{82}	Y_{83}	Y_{84}

where:

$Y'_i = (1/N_i) \times \Sigma_j\, y_{i,j}$

$S_i^2 = [1/(N_i - 1)] \times \Sigma_j\, (y_{i,j} - y_i')$

and i is the experiment number (outer array combination)

j is the trial number (inner array combination)

N_i is the total number of outer array combinations

After experiments, S/N ratios are calculated for the appropriate case: Note that these results are shown in Table 6.6.

For example, for variable or parameter 4, S/N ratios are consolidated one for level 1 and another for level 2.

$$\frac{S}{N}_{P4,L1} = SN1 + SN3 + SN5 + SN7$$

$$\frac{S}{N}_{P4,L2} = SN2 + SN4 + SN6 + SN8$$

TABLE 6.6

S/N Ratios

Experiment	Glucose	Peptone	$CaCl_2$	$FeSO_4$	NH_4NO_3	Tween 80	Vitamin B_{12}	S/N
1	1	1	1	1	1	1	1	SN1
2	1	1	1	2	2	2	2	SN2
3	1	2	2	1	1	2	2	SN3
4	1	2	2	2	2	1	1	SN4
5	2	1	2	1	2	1	2	SN5
6	2	1	2	2	1	2	1	SN6
7	2	2	1	1	2	2	1	SN7
8	2	2	1	2	1	1	2	SN8

TABLE 6.7

Consolidated S/N Ratios for Variables at Two Different Levels

Parameter Level	Glucose	Peptone	CaCl₂	FeSO₄	NH₄NO₃	Tween 80	Vitamin B₁₂
1	$SN_{V1,L1}$	$SN_{V2,L1}$	$SN_{V3,L1}$	$SN_{V4,L1}$	$SN_{V5,L1}$	$SN_{V6,L1}$	$SN_{V7,L1}$
2	$SN_{V1,L2}$	$SN_{V2,L2}$	$SN_{V3,L2}$	$SN_{V4,L2}$	$SN_{V5,L2}$	$SN_{V6,L2}$	$SN_{V7,L2}$
Delta	$SN_{V1,L2}$ $- SN_{V1,L1}$	$SN_{V2,L2}$ $- SN_{V2,L1}$	$SN_{V3,L2}$ $- SN_{V3,L1}$	$SN_{V4,L2}$ $- SN_{V4,L1}$	$SN_{V5,L2}$ $- SN_{V5,L1}$	$SN_{V6,L2}$ $- SN_{V6,L1}$	$SN_{V7,L2}$ $- SN_{V7,L1}$
Rank	2	3	1	4	6	7	5

Similar calculations for parameters 1, 2, 3, 5, 6, and 7 are shown in Table 6.7.

Table 6.7 gives the delta or the changes in the S/N ratios as the level of the parameter is changed. In addition, it shows the ranking of the parameter based on sensitivity to change and the noise.

6.8.1 Application of S/N ratio

The S/N ratio is applied to select the best set of parameter settings.[3]

Example 6.2

Calculate S/N ratio for the data in Table 6.8.

Solution:
Using Equation 6.1, one can calculate S/N ratio for individual runs. Table 6.9 summarizes the values of S/N ratio.

TABLE 6.8

Experimental Runs with Responses

Run	Chitin	Peptone	NH₄NO₃	FeSO₄	ZnCl₂	pH	Temperature	Response
				Variables with Their Levels				
1	1	1	1	1	1	1	1	0.94
2	1	1	1	2	2	2	2	1.20
3	1	2	2	1	1	2	2	1.07
4	1	2	2	2	2	1	1	1.03
5	2	1	2	1	2	1	2	1.10
6	2	1	2	2	1	2	1	1.10
7	2	2	1	1	2	2	1	1.14
8	2	2	1	2	1	1	2	1.08

TABLE 6.9

Values of S/N Ratio

Run Number	S/N Ratio
1	22.37091788
2	24.49198573
3	23.49603636
4	23.1651053
5	23.73621451
6	23.73621451
7	24.04645784
8	23.57683592

6.9 Orthogonal Array

Orthogonal array is the basis of Taguchi's design. It uses the concept of orthogonal[1] vectors to model the system. The orthogonal vectors are the primary constituents of the Taguchi design where the system to be optimized can be characterized by multiple levels and multiple factors.

The term *orthogonal array* has various meanings and interpretation. The term *orthogonal* simply means at right angles. This comes from its usage in vector algebra where two vectors are said to be orthogonal if their dot product is zero, indicating that geometrically they are perpendicular to each other. This means that the two vectors under consideration are independent to each other or they do not have any components along the direction of the other vector. This will not really make a big impact when we read about Taguchi's design as the various arrays that we will consider for a single problem will be orthogonal and of same length. It is better to know this as it may lead to avoiding errors and better manual debugging of solutions when practical problems are encountered.

This is a table of arrangement, having entries of finite set of symbols, namely, 1, 2, and so on. The following is the arrangement:

$$s - (l, v, i)$$

where:
s is the strength
l is the number of levels
v is the number of variables
i is the index (number of repeat)

TABLE 6.10

Orthogonal Array for Four
Variables at Three Levels

1	1	1	1
1	2	2	2
1	3	3	3
2	1	2	3
2	2	3	1
2	3	1	2
3	1	3	2
3	2	1	3
3	3	2	1

A simple array means no repetitive rows. For example, 2 – (3, 4, 1) orthogonal array is described by the following values:

2 = strength
3 = number of levels
4 = number of variables
1 = index, that is, number of repeat. Table 6.10 gives the arrangement.

Orthogonal arrays are always represented as *L-n*,[1] where *n* is a number. The letter *L* is used to represent Latin squares. The number of rows is *n* in the experimental design table. It is better to consider a simple example.

1. *L*-4 array: This particular orthogonal array has four rows or in other terms, when we design experiments for a particular system with this array, we will perform four different experiments to analyze that system.
2. Similarly, an *L*-8 array represents an orthogonal array, which has eight rows or when such a system is worked on, we have to perform eight experiments before we can understand the system.

Such a notation is not complete by itself. Take the first example, for instance; clearly, it never specifies the number of factors involved or the number of variables involved in the system under consideration. Therefore, to complete the representation, we use a *level* raised to the power of an *exponent* notation before our previously mentioned way of writing an array.

Assume that three major factors in a system at two different levels need to be controlled.

How can one represent the orthogonal arrays for this problem?

It is (2^3) *L*-4. The idea of why we chose an *L*-4 array will be explained later. Let us just go ahead with this for a better understanding of the representation.

The number 2 (or the base) represents the number of levels in which the system will be experimented on. Therefore, this indicates that the system will have two different levels of experimentation. The number 3, which is the exponent, shows the number of factors or variables for experimentation. This is actually the number of columns that are involved in the design matrix. We can further determine the number of possible combinations for this system by calculating the value of $2^3 = 8$. (base^{exponent}). This is nothing but the number of distinct (combinations of numbers) experiments that we can perform with this system. Alternatively, it represents the number of experiments we would have to perform in order to complete a factorial design of experiments for the same system (the most extensive and complete design).

In summary, these are the major points to look for while representing an orthogonal array for design using Taguchi's concept.[1]

- The number of levels of experimentation is the base and the number of factors of experimentation is the exponent of the first term represented within *parenthesis* [()] as (levels^{factors}).
- The term L is used to represent Latin squares.
- n is the number of experiments to be conducted for studying the system n follows the L with a -.

Taguchi's experimental orthogonal arrays[1] are L-4, L-8, L-12, L-16, and L-32. Figure 6.3 gives a schematic representation of the array.

Above representation of the array $s - (l, v, i)$ is also written as L_4 (2^3) (*cf.* Table 6.11, which means that

Number of experiments (N) = 4 entered in row

Number of levels of variables = 2 entered in different columns

Number of variables (K) = 3 entered in columns

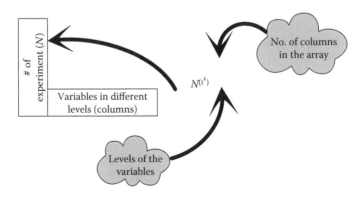

FIGURE 6.3
Short representation of orthogonal array.

TABLE 6.11

Representation of L_4 (2^3) or (2^3) L-4

Experimental Run Numbers	Variables		
	pH	Temperature	Agitation Rate
1	1	1	1
2	1	2	2
3	2	1	2
4	2	2	1

In an orthogonal array, the number of experiments is less than the number of experiments in a full factorial design. In this example, with the full factorial form, the numbers of experiments are $2^3 = 8$ (Figure 3.18). Thus, L-4 orthogonal array represents a 2^{3-1} fractional factorial design with reference to Tables 6.1 and 6.11. On the other hand, for Table 6.3, it is $3^{4-2} = 9$ possible combinations.

Tables 6.4 and 6.11 have the following features.

- Variables are coded as numbers, namely, 1, 2, 3, and so on.
- Rows (vertical entries) refer to different numbers of experiments.
- Columns (horizontal entries) represent variables at different levels. They are balanced, that is,
  - Equal numbers of levels of variables are present in a column.
  - In any two columns, the levels of combinations of variables appear in equal number.

For Table 6.10, all columns contain three 1-, three 2-, and three 3-levels, that is, balanced. Between any two columns one each of 1 1, 1 2, 1 3, 2 1, 2 2, 2 3, 3 1, 3 2, 3 3 combinations can be found.

- The first row has all at lower levels.
- There is no row at all at higher levels.

6.10 Taguchi's Method

Figure 6.4 describes Taguchi's method.

Example 6.3

In Table 6.12, the assumption is that the design of experimental runs is as per Taguchi's design. A secondary metabolite produced in a batch

Identification of main function, other effects, and ways of failure
↓
Identification of noise variables, conditions of experiments, and quality
↓
Identification of objective function (response) for optimization
↓
Identification of factors with their levels to be controlled
↓
Selection of orthogonal array, that is, balanced fractional factorial design, for the selection of experimental design matrix containing inner array to study the effects of design parameters and outer array for considering noise factors possibly affecting the performance of product/process quality.
↓
Performing the experiments as per the design matrix
↓
Analysis of data obtained from experiments and prediction of optimal levels
↓
Experimental verification of optimal value obtained in previous stage
↓
Planning the future

FIGURE 6.4
Schematic representation of Taguchi's method.

TABLE 6.12

Production of a Secondary Metabolite in a Controlled Batch Bioreactor

	Variables with Their Levels							Response (Secondary
Run	A	B	C	D	E	F	G	Metabolite, kg/m³)
1	1	1	1	1	1	1	1	0.94
2	1	1	1	2	2	2	2	1.20
3	1	2	2	1	1	2	2	1.07
4	1	2	2	2	2	1	1	1.03
5	2	1	2	1	2	1	2	1.10
6	2	1	2	2	1	2	1	1.10
7	2	2	1	1	2	2	1	1.14
8	2	2	1	2	1	1	2	1.08

bioreactor at controlled temperature and pH. Suitable inoculums level of proper age added to the medium. All runs performed in the same reactor of 3 m³ working volume. It was not possible to run another similar bioreactor, which could serve as the control.

Justify the statement by verifying the orthogonal criteria. Is it possible to represent the above plan into inner layer and outer layer strategy of Taguchi's design?

Solution:
Properties of Taguchi's orthogonal array are as follows:

1. The table of arrangement has finite set of symbols of 1 and 2.
2. There is no repeat row.
3. First row has all at 1-level. This example satisfies the criterion.
4. There is no row with all 2-levels. This is also true in this case.
5. All columns contain four 1-level and four 2-level. It is balanced array.
6. Between any two columns, there is two each of 1 1, 1 2, 2 1, 2 2 combinations.
7. The arrangement represents L_8 (2^7). In full factorial design, 2^7 experiments are necessary. Thus, L-8 orthogonal array represents 2^{7-4} fractional factorial design.

As per Taguchi's design, there are two layers: controllable variables and noise. In this array, all of the variables are controllable variables; no noise variable is present in this case. Alternatively, one can convert some the controllable variables into noise variable. This will give the arrangement for both layers.

Example 6.4

How can one calculate sum square (SS) and sum square total (SST) from the experimental response obtained in experiments carried out as per Taguchi's design plan?

Solution:
Theory
The relative influence of variables, for example, the variables indicated in Examples 6.1 and 6.2, may be calculated in the following way.[1]
The relative influence of each parameter on the production of a component in a biological reaction or any such response is described by the following equations.

$$SS = \frac{\sum_{i=1}^{n} (\text{Sum of yields in the } i \text{ level of one factor})^2}{\text{Number of yields in the } i \text{ level of that factor}}$$

$$- \frac{(\text{Total sum of yields})^2}{\text{Total number of yields}}$$

Percent SST = (SS/SST) × 100
Calculation is for sum of responses for individual level for a particular experiment and for a particular variable.
This will be easy to understand from Table 6.13.

TABLE 6.13

Data to Calculate SS and Percentage of SST

Run	Chitin	Peptone	NH_4NO_3	$FeSO_4$	$ZnCl_2$	pH	Temperature	Enzyme Activity
1	1	1	1	1	1	1	1	R_1
2	1	1	1	2	2	2	2	R_2
3	1	2	2	1	1	2	2	R_3
4	1	2	2	2	2	1	1	R_4
5	2	1	2	1	2	1	2	R_5
6	2	1	2	2	1	2	1	R_6
7	2	2	1	1	2	2	1	R_7
8	2	2	1	2	1	1	2	R_8

From Table 6.13, for variable A:

Total responses at level 1 = $R_1 + R_2 + R_3 + R_4 = R_{A1}$
Mean responses at level 1 = $(R_1 + R_2 + R_3 + R_4)/4 =$
Total responses at level 2 = $R_5 + R_6 + R_7 + R_8 = R_{A2}$
Mean responses at level 2 = $(R_5 + R_6 + R_7 + R_8)/4 =$
Therefore, SS for A variable = $SS_A = (R_{A1})^2 + (R_{A2})^2/4 - (R_{A1} + R_{A2})^2/8$
Sum square total = $SS_A + SS_B + SS_C + SS_D + SS_E + SS_F + SS_G$

Value of percentage of SST for each variable will suggest which variable is more important for further study.

6.11 ANOVA for Optimization of Experimental Parameters Using a Taguchi Design of Experiment

The design of experiment, the conduct of experiment, and the record of results are not meaningful for the correct decision on the system, unless statistical analyses have been made in this regard; ANOVA is necessary.[1] Two different ANOVA analyses are simple and advanced ANOVA.

6.11.1 Simple ANOVA

The simple ANOVA helps one find out information about

- The main effects of the parameters
- The parameter conditions to achieve optimal production
- A measure of the performance or the reliability one can expect at the before mentioned optimal conditions

6.11.2 Advanced ANOVA

An advanced ANOVA helps to find out the following:

- A confidence interval for the predicted optimal performance
- A measure for the relative influence of the parameters on the performance
- Influence of errors that arise in the experiments
- Tests of significance for parameters

6.11.3 Systematic ANOVA Analysis

Figure 6.5 gives the details of systematic ANOVA analysis.

Example 6.5

A biological system is studied by Taguchi's design. Results are shown in Table 6.14. The aim of this study is to perform ANOVA and draw meaningful conclusions about the system. The system has seven variables, with each variable being able to take up two different levels. The orthogonal array for such a system meets Taguchi's design.

Defining a new term called the *total T*. As the name suggests T is the sum of all the responses for all the systems. In this case, sum of $Y1, Y2, ... Y8$. Thus,

$$T = \sum Y_i$$

Main effects term for each individual parameter for each level is as follows.

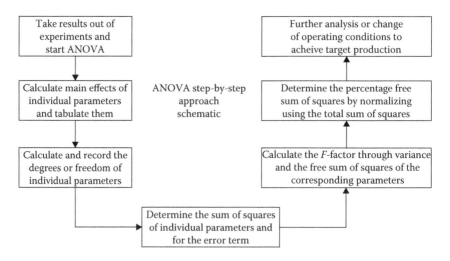

FIGURE 6.5
Systematic ANOVA.

TABLE 6.14

L-8 Orthogonal Array

Experimental Run Numbers	Chitin	Peptone	NH$_4$NO$_3$	FeSO$_4$	ZnCl$_2$	pH	Temperature	Enzyme (Y)
1	1	1	1	1	1	1	1	E1
2	1	1	1	2	2	2	2	E2
3	1	2	2	1	1	2	2	E3
4	1	2	2	2	2	1	1	E4
5	2	1	2	1	2	1	2	E5
6	2	1	2	2	1	2	1	E6
7	2	2	1	1	2	2	1	E7
8	2	2	1	2	1	1	2	E8

A_i is the sum of all responses for experiments conducted at the *i*th level of a parameter.

The average effect of a parameter = the the average of responses for experiments conducted at level *i*.

$$<A_i> = \frac{A_i}{n_i}$$

Degrees of freedom for parameter *A* = number of levels of parameter *A* − 1.
Degrees of freedom of a column = number of levels of the column − 1.
Degrees of freedom of an experiment = total number of results of all trails − 1.
Degrees of freedom of an array = sum of degrees of freedoms of all the columns.

The total sum of squares is calculated as the sum of the squared values of all the differences of individual data points from the mean of the responses. Mathematically, it can be shown as follows:

$$S_T = \Sigma(Y_i - Y')^2$$

where:
   Y' is the mean of all the responses

Considering that the sum of all responses is quantified as *T*, it would be meaningful to write the above expression in terms of *T*. So expanding the above equation, we get

$$S_T = \Sigma(Y_i^2 + Y'^2 - 2Y_iY'^2)$$

But we know that $Y' = T/N$, so using this and taking the summation we get,

$$S_T = \left(\Sigma Y_i^2\right) - T^2 / N$$

Similarly, the sum of squares for the individual parameters is as follows:

$$S_A = \left(\frac{A_1^2}{N_{A1}} \right) + \left(\frac{A_2}{N_{A2}} \right) + \cdots - CF$$

where:
  CF is the correction factor, $CF = T^2/N$

Similar definitions can be made for each individual parameter. The sum of squares error is

$$S_E = S_T - (S_A + S_B + S_C \ldots)$$

It can also take up a value of zero. The next step is to determine the variances of the individual parameters:

$$V_i = \frac{S_i}{f_i}$$

where:
  f_i is the degrees of freedom for the ith parameter

It has been observed that $F_i = V_i/V_e$ follows an F-distribution with (f_i, f_e) degrees of freedom. What good can the variance do to us if we are not able to bring it to an interpretable F-value? $F_i = V_i/V_e$; in a case where the error contributions cannot be determined, then the F-value cannot be decided as it has no realistic value. Once we have calculated the F-value, we can use an F-table and decide the confidence level for our interpretation. One has to be careful to use the right degrees of freedom and the correct F-value. The degrees of freedom of both the parameter and of the error are to be used in that order.

Finally, the percent contribution is necessary. This is the amount or the percentage of the overall change in values that is being contributed by the given factor. For this purpose, one cannot just take the sum of square values of the individual parameters, as it can be misleading since the error factors have to be accounted. It is necessary to create a new corrected pure sum of squares for each term and then use it for further calculations.

This is done as follows:

$$S_i' = S_i - \left(V_E f_i \right)$$

This is followed for all the individual parameters to give us pure sum of squares for all the individual parameters.

Then calculate the percent contributions by the different parameters and the remaining from 100 will give the contribution of errors.

$$P_i = \frac{S_i'}{S_T}$$

$$P_e = 100 - (P_A + P_B + P_C)$$

Example 6.6

Consider a biological system like a continuous flow stirred tank reactor, which is cultivating a bacterium that is needed for us in large quantities to get some useful product. We know that a number of different factors or parameters determine the system's performance. The system has been played with, and the identified key parameters are the pressure in the reactor, amount of carbon in the medium that is providing the nutrition for the bacteria to grow, and the temperature of operation of the reactor. Also, let us assume that we have flexibility in selecting two different conditions for each factor. Each parameter has two levels. Use Taguchi's experimental plan to decide the system.

Solution:

Therefore, we have three variables and we need to perform a Taguchi's design on this system. Since it is a real reactor system, the cost involved with every experiment is very high. Thus, Taguchi's design is the right way to go as other designs lead to very large amount of resource and time consumption through multiple experiments. The experimental design is as per Table 6.15.

From the experiments, responses are obtained from the system. In this case, it is the amount of the product formed – in terms of the concentration. The responses are shown in Table 6.16.

The next step following the experiment is to analyze the results using an ANOVA analysis.

The total of response T becomes

$$T = \Sigma Y_i = 1 + 2 + 3 + 4 = 10$$

TABLE 6.15

Experimental Plan

Experiment Number	Variable A – Pressure in the Reactor	Variable B – Carbon Concentration	Variable C – Temperature Maintained in the Reactor
1	1	1	1
2	1	2	2
3	2	1	2
4	2	2	1

TABLE 6.16

Experimental Conditions and Responses

Experiment Number	Variable A – Pressure in the Reactor	Variable B – Carbon Concentration	Variable C – Temperature Maintained in the Reactor	Responses (y)
1	1	1	1	1
2	1	2	2	2
3	2	1	2	3
4	2	2	1	4

The next step is to calculate the main effects of the individual parameters at different levels:

A_1 = Experiment 1 + Experiment 2 = 1 + 2 = 3
A_2 = Experiment 3 + Experiment 4 = 3 + 4 = 7
B_1 = Experiment 1 + Experiment 3 = 1 + 3 = 4
B_2 = Experiment 2 + Experiment 4 = 2 + 4 = 6
C_1 = Experiment 1 + Experiment 4 = 1 + 4 = 5
C_2 = Experiment 2 + Experiment 3 = 2 + 3 = 5

Now, we calculate the change in response when the parameter level is increased or the main effect as follows:

For parameter $A = A_2 - A_1 = 7 - 3 = 4$
Parameter $B = B_2 - B_1 = 6 - 4 = 2$
Parameter $C = C_2 - C_1 = 5 - 5 = 0$

The next step is to determine the degrees of freedom for this system. We use the formula as mentioned before and calculate the degrees of freedom for all three parameters.

Degrees of freedom of parameter A = number of levels of $A - 1 = 2 - 1 = 1$
Degrees of freedom of parameter B = number of levels of $B - 1 = 2 - 1 = 1$
Degrees of freedom of parameter C = number of levels of $C - 1 = 2 - 1 = 1$
Total degrees of freedom of experiment = total trials $- 1 = 4 - 1 = 3$

Therefore, the degrees of freedom of errors = total degrees of freedom − degrees of freedom of A − degrees of freedom of B − degrees of freedom of $C = 3 - 1 - 1 - 1 = 0$.

Thus, it is to be noted that, in this case, the degrees of freedom for the error is zero. We did not carry out any repeat experiments, which means that we have no means of deciding if any change in the response is due to the error or not. Hence, there is no freedom in calculating the error effects.

One can come to a preliminary conclusion that the response seems to be very highly dependent on parameter A and that both parameters A and B are such that an increase in the levels of the parameters leads to an increase in the response. Parameter C does not seem to change anything. No quantitative inference can be drawn with much confidence, as there is no information regarding the interaction effects. Interaction effects can make any of the above conclusions invalid. The above conclusions are good for an approximate overall idea about the system and to get a feel for how the response of the system is affected by the change in the levels of different parameters.

Let us look at the advanced ANOVA analysis.

The sum of squares for each of the parameters is calculated as given below.

$$S_A = \left(\frac{A_1^2}{N_{A1}} \right) + \left(\frac{A_2^2}{N_{A2}} \right) - CF$$

where:
CF is the correction factor $= T^2/N$
Thus $S_A = 3^2/2 + 7^2/2 - 10^2/4 = 4.5 + 24.5 - 25 = 4$

Similarly, one can calculate S_B, S_C, and S_T.

$$S_B = \left(\frac{B_1^2}{N_{B1}} \right) + \left(\frac{B_2^2}{N_{B2}} \right) - CF$$

where:
CF is the correction factor $= T^2/N$
Thus, $S_B = 4^2/2 + 6^2/2 - 10^2/4 = 8 + 18 - 25 = 1$

$$S_C = (C_1^2/N_{C1}) + (C_2^2/N_{C2}) - CF$$

Thus, $S_C = 5^2/2 + 5^2/2 - 10^2/4 = 12.5 + 12.5 - 25 = 0$

$$S_T = \Sigma Y_i^2 - CF$$

Thus $S_T = (1^2+2^2+3^2+4^2) - 10^2/4 = 1 + 4 + 9 + 16 - 25 = 5$.

It can be observed that $S_T = S_A+S_B+S_C$ as $5 = 4+1+0$.

Therefore, one can conclude again here that $S_E = 0$ as $S_E = S_T - (S_A+S_B+S_C)$.

We already know the variance factor for a parameter is the sum squares for that parameter divided by the degrees of freedom of that parameter. Since the degree of freedom for all the parameters is 1, the S_i is the same as V_i. For the error terms, both S_E and the degrees of freedom (f_e) are zero. V_e is indeterminate. Since V_e cannot be decided, we cannot calculate the F factors for the individual parameters.

The aim is to calculate the percentage contributions by the individual parameters. For this, one has to determine the pure sum of squares of the individual factors by discounting the effect of the errors that creep in during the experiment. In the experimental design, since there are no repeat experiments, the pure sum of squares of the parameters is the same as that of sum of squares of individual parameters.

$$S_i' = S_i - \left(f_i V_e \right)$$

Since V_e does not exist, in our case $S_i' = S_i$.

Using the formula

$$P_i = \frac{S_i'}{S_T}$$

One can calculate the percentage contributions of each of the above parameters and make necessary decision to the individual parameters to achieve a target production or one can also pool the variables with very little contributions and then go for further vigorous experiments.

$$P_A = \left(\frac{S_A}{S_T} \right) \times 100 = 100 \times \frac{4}{5} = 80\%$$

$$P_B = \left(\frac{S_B}{S_T} \right) \times 100 = 100 \times \frac{1}{5} = 20\%$$

$$P_C = \left(\frac{S_C}{S_T} \right) \times 100 = 100 \times \frac{0}{5} = 0\%$$

From this, one can conclude that factor C does not influence the response to a detectable extent. A change in factor A leads to about four times the change in response than that obtained by same change in factor B. For implementation of changes to the system, after this study, one needs to observe how far the current production quality is away from the target quality. Then factor A needs to change and control production until it reaches a value very close to the target value. Once it reaches this value, the change in small values of A may give too much of a change in the response. It shoots up on the other side of the target value. In such a situation, one uses the other control, which is parameter B to fine-tune the final response value to achieve the target value. This is because the response is not as sensitive to parameter B as it is to parameter A.

6.12 Limitation in Taguchi's Design[1]

In Taguchi's design, the S/N ratio is a function of the following components:

- Factors influencing process mean and variance
- Factors influencing on process mean

Hence, in this design,

- The S/N ratio does not allow separate treatment for the roles of two different natures of factors.
- In addition, this uses cumbersome crossed array design.

In the literature, the following points are the probable doubts for the use of this design:[6]

- To develop system insensitive to environmental factors.
- The designed product is insensitive to variation caused by the composition of the components.
- The designed process may be closed to recommended product; however, it may be influenced by the process variables and nature of raw materials.
- Determined operating conditions are critical, which may vary due to influential factors.

6.13 Outcome of Taguchi's Design

Though there are difficulties in implementing the technique and sizeable limitations associated with it, it has been felt that the outcome of this important method of design is the following.[6]

- Acceptable in industrial sectors
- Successful integration of good experimental design practice into service and engineering
- Development of combined array designs and the response model approach

6.14 Application of Taguchi's Design

Taguchi's design considers separately the controllable and noise factors. In this case, the noise factors combines with controllable factors, which give the concept of a combined array design. Montgomery[6,7] suggested both controlled and noise factors should be coded uniformly having centre points plus upper and lower limits. As per his suggestion, the combined array assumes a first-order model (Equation 6.2).

$$\tilde{Y} = \beta_0 + \beta_1 x_1 + \beta_2 x_2 + \cdots + \beta_{12} x_1 x_2 + \psi_1 z_1 + \Phi_{11} x_1 z_1 + \Phi_{21} x_2 z_1 + \varepsilon \quad (6.2)$$

where:
  z_1 is the noise factor in coded form
  x_1 and x_2 are controllable coded factors
  Φ_{11} and Φ_{21} are regression coefficients
  ψ_1 is the coefficient associated with the noise factor
  ε is error

Now, in the combined array design, both controllable variable and noise variable are in a single experimental design. This avoids the inner and outer array rearrangements in Taguchi's design.

Exercises

6.1 Explain Taguchi's experimental design plan with a proper example from biological science.

6.2 Analyze the data in Table 6.2.1, which are obtained from Taguchi's robust design for the production of citric acid in a batch bioreactor.

6.3 Prepare orthogonal array as per Taguchi's design concept for the following configuration:

  1. $L_{12}(3^{11})$
  2. $L_8(2^7)$

6.4 What is the difference between the conventional experimental designs and Taguchi's ways of defining quality?

TABLE 6.2.1

Percent SST for Each Variable

Factor	Percentage of SST
Glucose (kg/m³)	15
K-phosphate (kg/m³)	0.80
FeCl₃ (kg/m³)	0.01
NaCl (kg/m³)	5.30
Agitator speed (rev/min)	45
Air flow rate (m³ air/m³ medium, min)	24
pH	34.5

TABLE 6.6.1

L-4 Array

Experimental Run Number	pH	Air Flow Rate	Agitation
1	1	1	1
2	1	2	2
3	2	1	2
4	2	2	1

6.5 What is an orthogonal array and what are its advantages?

6.6 Select the rows of an *L*4 Taguchi design array from Table 6.6.1. Prove that the individual rows are indeed orthogonal (Hint: treat each row as a vector and perform a dot product).

6.7 When will you use a the Taguchi design and when will you deem it not so useful?

6.8 Consider a biological system, which is being optimized for better production of a recombinant product. After a theoretical study, it has been identified that there are three major variables. Those variables that affect the output of the system are *A*, *B*, and *C*. Taguchi's design was performed and the responses were recorded in Table 6.8.1.

Now, having performed those experiments, determine the main effects of the individual variables and determine if any of these variables can be deemed insignificant at a contribution limit of 10%.

6.9 Try and design a Taguchi array for four-variable (use dummy variables if necessary) system with three levels for each individual variable and cross verify your answer by performing the dot product analysis by assuming that the individual rows are separate vectors.

6.10 You are a consultant who has been hired by a biochemical company to optimize its production levels. Due to intellectual property rights of the company, you do not have any information about the product or how it is produced, but you are given a brief lecture about the product under false names. So variables that affect the rate of

TABLE 6.8.1

Experimental Conditions with Response

Experimental Run Number	*A*	*B*	*C*	Response
1	1	1	1	5.4
2	1	2	2	6.2
3	2	1	2	3.7
4	2	2	1	8.9

TABLE 6.10.1

Rate of Production of *P* under Various Experimental Conditions

Experiment	*A*	*B*	*C*	*D*	*E*	*F*	*G*	Rate of Production of *P*
1	1	1	1	1	1	1	1	102.7
2	1	1	1	2	2	2	2	44.9
3	1	2	2	1	1	2	2	109.6
4	1	2	2	2	2	1	1	127.44
5	2	1	2	1	2	1	2	118.11
6	2	1	2	2	1	2	1	50.3
7	2	2	1	1	2	2	1	123.45
8	2	2	1	2	1	1	2	154.67

production of *P* are *A*, *B*, *C*, *D*, *E*, *F*, and *G*. As the consultant, you have reviewed the cases and have suggested that a Taguchi's design be performed and the results are in Table 6.10.1.

Now, in the next step in your analysis, you have to perform an ANOVA analysis and determine the *F*-factor used for statistical interpretations. Tabulate the *F*-values for each factor along with their degrees of freedom.

6.11 A biological system had to be optimized to yield a higher productivity. For this purpose, a set of experiments was performed and the results are shown in Table 6.11.1. Incidentally, the distinct experimentation points were not different from that used for Taguchi's experimental

TABLE 6.11.1

Experimental Plan with Response

Experiments	pH	Air Flow Rate	Agitation	Temperature	Glutamic Acid
1	1	1	1	1	0.45
2	1	2	2	2	1.23
3	1	3	3	3	1.10
4	2	1	2	3	1.15
5	3	3	2	3	0.78
6	2	2	3	1	0.83
7	2	2	3	1	0.99
8	2	3	1	2	0.91
9	3	1	3	2	0.89
10	1	1	1	1	0.94
11	3	2	1	1	1.04
12	1	1	1	1	1.12
13	3	3	2	3	0.62
14	3	1	3	2	1.19
15	1	2	2	2	0.75

design. Determine the *degrees of freedom* of each variable and the *degrees of freedom* of the entire experiment along with the *degrees of freedom* of the errors.

6.12 Consider the optimization of the growth of *Escherichia coli* in a reactor for industrial purposes. The growth of the bacteria can be influenced by multiple factors such as type of reactor, time of incubation, temperature of operation, pressure of operation, type of mixing, and speed of mixing impeller. Now, we have 11 random variables chosen for helping us optimize the performance of process. We are asked to determine the percentage contributions of each of the variables and pick the best five. (Not the simple ANOVA analysis and main effects but use the advanced ANOVA analysis – find the *F*-factor first and then go for percentage contributions.)

A Taguchi's design has been performed and the responses have been recorded in Table 6.12.1. Use the above table to solve the problem.

6.13 Consider a biological reactor that is optimizing the amount of biological substance, which is being produced due to the activity of a particular bacterium in the reactor. It has been identified that there are three major factors, which are manually controlled variables that allow us to manage the process while there are two factors that during normal operation we cannot control completely. Each of these (including the uncontrollable disturbances) can take up two different levels. A Taguchi's design of experiment was performed with separate outer and inner arrays for this system, considering Taguchi's defined concept of quality and the responses have been recorded in the Table 6.13.1. The objective of the product producer is to achieve a good production while also making sure that there is

TABLE 6.12.1

Experimental Plan with Response

Experiments	f1	f2	f3	f4	f5	f6	f7	f8	f9	f10	f11	Response
1	1	1	1	1	1	1	1	1	1	1	1	42.12
2	1	1	1	1	1	2	2	2	2	2	2	43.57
3	1	1	2	2	2	1	1	1	2	2	2	42.74
4	1	2	1	2	2	1	2	2	1	1	2	41.98
5	1	2	2	1	2	2	1	2	1	2	1	44.12
6	1	2	2	1	2	2	1	2	1	2	1	43.43
7	1	2	2	2	1	2	2	1	2	1	1	43.71
8	2	1	2	1	2	2	2	1	1	1	2	42.86
9	2	1	1	2	2	2	1	2	2	1	1	42.10
10	2	2	2	1	1	1	1	2	2	1	2	42.74
11	2	2	1	2	1	2	1	1	1	2	2	42.82
12	2	2	1	1	2	1	2	1	2	2	1	43.25

TABLE 6.13.1

Experimental Plan with Response

Experiment	pH	Temperature	Air Flow Rate	Response 0,0	0,1	1,0	1,1
1	1	1	1	1.02	1.04	1.09	0.99
2	1	2	2	1.19	1.23	1.22	1.27
3	2	1	2	0.85	0.82	0.84	0.79
4	2	2	1	1.42	1.46	1.39	1.44

not much change in the production values form its specified value due to disturbances. For this purpose, calculate the S/N ratio table for this system along with the percentage contributions and marginal contributions for each variable.

References

1. Roy RK (Ed.), *Design of Experiments Using Taguchi Approach*, John Wiley & Sons, New York, 2001.
2. Roy RK, *A Primer on the Taguchi Method*, Van Nostrand Reinhold, New York, 1990.
3. Bruns RE, Scarminio IS, and de Barros Neto B (Eds.), *Statistical Design – Chemometrics*, Elsevier, Amsterdam, the Netherlands, 2006.
4. Phadke MS (Ed.), *Quality Engineering Using Robust Design*, Prentice Hall, Englewood Cliffs, NJ, 1989.
5. Vuchkov IN and Boyadjieva LN, *Quality Improvement with Design of Experiments: A Response Surface Approach*. Kluwer Academic Publishers. Dordrecht, the Netherlands, 2001.
6. Montgomery DC (Ed.), *Design and Analysis of Experiments*, 7th edition, Wiley, India, pp. 1, 20, 486, and 493, 2009.
7. Hinkelmann K, *Design and Analysis of Experiments: Special Designs and Applications*, Vol. 3, John Wiley & Sons, Hoboken, NJ, 2012.

Further Reading

1. Rao RS, Kumar CG, Prakasham RS, and Hobbs PJ, The Taguchi methodology as a statistical tool for biotechnological applications: A critical appraisal, *Biotechnology Journal*, **3**, 510–523.
2. Dasu VV, Studies on production of griseofulvin by *Penicillium griseofulvum*, PhD Thesis, Indian Institute of Technology, Madras, India, 1999.
3. Phadke MS and Taguchi G, Selection of quality characteristics and S/N ratios for robust design, In conference record, *GLOBECOM87 Meeting*, IEEE, 1987.

7

Hybrid Experimental Design Based on a Genetic Algorithm

OBJECTIVE: A genetic algorithm takes care of different variables for searching optimal conditions. Hybrid experimental design shows better exploitation of biological systems.

7.1 Introduction

Studies on evolutionary systems by computer scientists with the concept of using evolution are one of the techniques of optimization to solve engineering problems. The idea was to evolve a population of candidate solutions to a given problem, using operators inspired by natural genetic variation and selection.

A genetic algorithm, as its name suggests, has close relations to genes. It is a search algorithm that mimics natural process of the survival of the fittest.[1] In every generation, a new set of offspring (strings) are created, which are derived from the information in the parent creatures or strings, taking more information from the fittest of the parental strings. The fitness criteria are set beforehand and are evaluated by the experimenter. Occasionally, a new part adds to the new generation string or modifies and the process is continued.

Unlike what it seems, the genetic algorithm is not just a random search algorithm[2] used to find the most efficient variable levels (experimental conditions), but it is a very efficient method. This exploits the historical information available from the previous generations of solutions to provide information about new experimental conditions that can possibly provide better results than the previously used experimental conditions. However, there is no guarantee in this regard. It is very much possible that a new generation of strings (points) show lesser fitness than their parents do, but in due course of time, after enough iterations (enough generations of experiments or fitness evaluations), following the genetic algorithm will provide better points than the old ones.[2]

Although genetic algorithms were based on the concept of evolution introduced by Darwin, Holland et al.[1] formally introduced this algorithm.

Holland et al.'s concept has been twofold[1]:

1. To abstract and to explain rigorously the adaptive and evolutionary process of natural systems
2. To design artificial systems software that retains the important mechanisms of natural systems

Their research has led to important discoveries in both natural and artificial systems.

7.2 Need for Search Algorithms

In most industrial processes, there is a need to find and operate equipment or set the control variables at a particular level to have improved production of the required products. There are continuous and extensive conditions to be maintained for each variable. It is very difficult to meet the correct conditions.[1]

Therefore, out of the enormous available possible solutions, one has to consider the best one; search algorithms do this.[3] The genetic algorithm is one of the search algorithms that can help solve the above problem.

Such search problems can often benefit from an effective use of parallelism, in which many different possibilities explored simultaneously in an efficient way.

Biological evolution is an appealing source of inspiration for addressing such problems.[1] Evolution is, in fact, a method of searching among an enormous number of possibilities for *solutions*. In biology, the enormous set of possibilities is the set of possible genetic sequences. The desired solutions are highly fit organisms, which are able to survive and reproduce in their environments. Evolution can also be thought as a method of designing innovative solutions to complex problems.[4]

To search for a solution in the face of changed conditions is precisely required for adaptive computer programs.[1] Furthermore, the evolution is a massively parallel search method.

7.3 Method

The natural process of evolution directly leads to the genetic algorithm when viewed form an engineering perspective. In the process of evolution, the genetic information – better known, as *deoxyribonucleic acid* (DNA) is responsible for all the characteristics and behaviour of an organism that shows variation from individual to individual based on the difference in the DNA of the parental organisms.[5] The changes occur by the natural processes of mutation and crossover. Similar to this, in the genetic algorithm from a parental set of solutions, a new generation of solutions is obtained, and changes are in the form of mutations and crossover.

In the evolution process, the mutation that leads to better or fitter characteristic to the organism survives and becomes the dominant source for the generation of future organisms. Similarly, one evaluates the fitness function for all the parental strings and picks the strings to produce the offspring based on this result. The string that has higher fitness evaluated has a higher probability of selection to produce the offspring or the next generation of solutions.

7.3.1 Procedure

7.3.1.1 Theory

In this algorithm, individuals/creatures termed as *population* of candidate solutions for optimization studies evolves better solutions. A set of properties for each solution required for mutation and alteration. Classically, solutions are strings of zero and one (binary nomenclature), but other codes are also possible.

Steps

1. The evolution usually starts from a population of randomly generated individuals by an iterative process, with the population in individual iteration called a *generation*.

2. In each generation, every individual in the population is evaluated for fitness. Fitness is usually the value of the objective function in the optimization problem, which is being solved. The more fit individuals are stochastically[1] selected from the current population.

3. Each individual's genome is modified (recombined and possibly randomly mutated) to form a new generation of solutions to the given problem.

4. The new solutions are in the next iteration of the algorithm as the base case (or parent creatures).

5. The algorithm terminates either when a set number of generations attained or when a satisfactory fitness level is achieved for the population.

7.3.1.2 Candidate Defined

A standard representation of each candidate solution is as an array of bits.[1] Arrays of other types and structures can be used in essentially the same way. The main property that makes these genetic representations convenient is that their parts easily align due to their fixed size, which facilitates simple crossover operations. Variable length representations may also be used, but crossover implementation is more complex in this case.

7.3.1.3 Typical Genetic Algorithm

Any genetic algorithm requires solution domain and the fitness function to get the solution domain.[6]

Section 7.3.2 discusses elaborately on genetic representation and fitness function.

7.3.2 Genetic Representation of the Solution Domain

Initially, many individual solutions are (usually) randomly[7] generated to form an initial population. The population size depends on the nature of the problem, but typically contains several hundreds or thousands of possible solutions. Traditionally, the generated population is random, allowing the entire range of possible solutions (the search space) to be possibilities for parent solutions. The solutions may be *seeded* quite often for the availability of likely optimal solution.[1]

7.3.3 Fitness Function for Evaluating the Solution Domain

During each successive generation, a proportion of the existing population selects to breed a new generation. Individual solutions selected through a fitness measurement process, where fitter solutions (as measured by a fitness function) are more likely to be selected.[8] Certain selection methods rate the fitness of each solution and preferentially select the best solutions. Other methods rate only a random sample of the population, as the former process may be very time-consuming.[1]

The fitness function operates independent of the type of genetic representation and measures the quality of the represented solution. The fitness function is problem dependent. In some problems, it is hard or even impossible to define the fitness expression. In these cases, a simulation may be used to determine the fitness function value.[1]

The next step is to generate a second generation (population) of solutions from those selected through genetic operators: crossover (also called *recombination*) and/or mutation.[1]

For each new solution to be produced, a pair of *parent* solutions is selected for breeding from the pool selected previously.[1] By producing a *next progeny* solution using the above methods of crossover and mutation, a new solution is created, which typically shares many of the characteristics of its *parent population*. Each new progeny comes from the selected new parents. This process continues until it generates a new population of solutions of an appropriate size. Although, reproduction methods based on the use of two parents is *biologically inspired*, some research suggests that more than two *parents* generate higher quality chromosomes.[2]

These processes ultimately results in the next generation population of chromosomes that are different from the initial generation.[1] Generally, the average fitness will be increased by this procedure for the population, since only the best organisms from the first generation are selected for breeding, along with a small proportion of less fit solutions. These less fit solutions ensure genetic diversity within the genetic pool of the parents and therefore ensure the genetic diversity of the subsequent generation of solutions.[4]

Although crossover and mutation are known as the main genetic operators, it is possible to use other operators such as regrouping, colonization-extinction, or migration in genetic algorithms.[4]

It is worth tuning parameters such as the mutation probability, crossover probability, and population size to find suitable settings for the problem class chosen. A very small mutation rate may lead to a genetic drift (which is non-ergodic[1] in nature). A recombination rate that is too high may lead to premature convergence of the genetic algorithm.[6] High mutation rate causes loss of good solutions unless specific selection required.[7] No appropriately defined guidance is available for suitable boundary values of these parameters for selection through experiments.

This process of generation continues until a termination condition is reached. Common terminating conditions are as follows:[1]

- A solution must satisfy the minimum fitness criteria.
- Fixed number of generations (iterations of the algorithm application) must be attained.
- A budget, that is, computation of time/money, should be matched.
- After attaining highest-ranking solution's fitness, later iterations do not produce better results.
- An inspection by the experimenter.
- Combinations of the above points.

7.4 Terminologies

7.4.1 Genetic Operators

7.4.1.1 Mutation

Mutation is a genetic operator used to maintain genetic diversity from one generation of a population of genetic algorithm chromosomes to the next.[1] It is analogous to biological mutation. Mutation alters one or more gene values in a chromosome from its initial state. In mutation, the solution may change entirely from the previous solution. Hence genetic algorithm (GA) can come to better solution by using mutation.[6] Mutation occurs during evolution according to a user-definable mutation probability. This probability should be set low. If it is set too high, the search will turn into a primitive random search.[7]

A common method of implementing the mutation operator involves generating a random variable for each bit in a sequence.[1] This random variable tells whether or not a particular bit will be modified. This mutation procedure, based on the biological point mutation, is called *single-point mutation*. Other types are inversion and floating-point mutations.

7.4.1.1.1 *Purpose of Mutation*

Mutation in GAs takes care of diversity to our procedure, which can help us reach the final optimal solution faster.[4] Mutations should allow the algorithm to avoid local minima by preventing the population of chromosomes from becoming too similar to each other, thus slowing or even stopping evolution. This reasoning also explains the fact that most GA systems avoid only taking the fittest of the population in generating the next generation of solutions but rather a random (or semi-random) selection with a weighting towards those that are fitter.

7.4.1.1.2 *Types of Mutation Operations*

There are different types of mutation operations that can be combined with the genetic algorithm. The major ones are listed as follows[2]:

- *Flip bit*: This mutation operator takes the chosen genome and inverts the bits. (i.e., if the genome bit is 1, it is changed to 0 and vice versa).
- *Boundary*: This mutation operator replaces the genome with either lower or upper bound randomly. This is specifically useful when the genome is not only 0s and 1s.
- *Uniform*: This operator replaces the value of the chosen gene with a uniform random value selected between the user-specified upper and lower bounds for that gene.
- *Non-uniform*: With this operator, probability of mutation will go to 0. In the early stages of the evolution, this avoids stagnation of population. Solutions in later stages of evolution are adjusted.
- *Gaussian*: A normalized Gaussian distributed random value is added to the chosen gene. If it is outside the boundary values for that gene, the new gene value is mapped.
- Integer and float genes are the characteristics of uniform, non-uniform, and Gaussian cases.
- *Bit string mutation*: The mutation of bit strings ensue through bit flips at random positions.

7.4.1.2 Crossover

In genetic algorithm, the term *crossover* refers to the process of producing a generation of offspring from a given generation.[1] It is the process of reproduction. In this case, solutions for the next generation come from solutions of the previous generation with proper selection and mixing. It is very similar to the biological process of crossover that happens during reproduction, where two genes combine to produce two new genes by mixing with each other.

7.4.1.2.1 Types of Crossover

The different types of crossover operations are summarized below[2]:

1. *One-point crossover*: The number of crossover points as per this method is one. A particular point is chosen from each gene after selecting two genes (genes here refer to solutions to the problem in hand) initially. The two genes are cut and swapped, which combine to produce two new solutions. The new strings or genes produced are termed as the *next generation of solutions*.

2. *Two-points crossover*: Two points are selected in the parent organism strings. The portion of the solution between the two points swapped in this method to produce new generation of solutions. Two new solutions are formed from two starting solutions.

3. *Cut and splice crossover*: This crossover variant uses the cut-and-splice approach, which results in a new set of solutions having unequal lengths. This happens because instead of choosing the same point in both the strings for crossover, we chose different points in different strings; this leads to unequal lengths when they are swapped and combined.

4. *Uniform crossover*: This method of crossover is known to use a fixed mixing ratio between the two parents. This has the unique property of allowing the parent genes or strings to contribute the gene level rather than the segment level. The mixing ratio that determines how much of genes come from a particular gene; the rest come from the other parent gene. A mixing ratio of 0.5 means equal contribution from both the parents towards the new generation solution. Although the contribution is fixed, the crossover points varies, as the selection is random. Uniform crossover evaluates each bit in the parent strings for exchange with a probability of 0.5.

5. *Three-parent crossover*: Unlike any of the methods mentioned above, this method of crossover is the only method that uses three parent strings to come up with the new generation strings. This is a random selection. Bits from one of the parents compare with corresponding bits form the next parent. If they are same, the selected bit is a part of the offspring; otherwise, the bit from the third parent takes directly for the offspring.

7.5 Limitations of Genetic Algorithm

Although it is a very efficient and a widely used method for solving engineering problems, the usage of genetic algorithm has a few limitations, which are mainly due to the practical constraints that come up while using the algorithm in actual practice.
The limitations are as follows:

1. It involves the repeated evaluation of the fitness functions that have to be maximized or minimized.[6] For example, the experimenter wants to maximize the amount of a particular type of bacteria that produces a useful intracellular product. The aim is to optimize the cell concentration in the reactor under different medium conditions. This is a well-known industrial problem. Large array of starting conditions are available for this experiment. The beginning starts with an initial guess and starts with a set of proposed solutions with proper evaluation of their fitness values. Here, the fitness function is the amount of cells that is produced under different concentrations of the medium. The evaluation of this fitness function for every proposed generation of solutions is required as per the genetic algorithm. It is necessary to perform multiple evaluations of the fitness function simulations prior to the consideration of next generation. In some cases, one may end up getting solutions that do not perform as good as the previous generation. Hence, all the effort for doing experimental or computational work may be of no use. Thus, it is highly tedious depending on the fitness function and its ease of determination. Therefore, it can be highly effort consuming as well as time- and resource-consuming process.

2. The better search solution using the genetic algorithm is only in comparison with the other solutions.[2] So, the particular termination criteria for the program or particular scenario is not always clear, since it is not always possible to search for the best possible solution without trying out every possible experimental point in the domain. The genetic algorithm keeps proceeding until meaningful stop criterion, which is especially for the cases with continuous variables. In the case of discrete variables, it will stop eventually due to the finite numbers. Even if the best solution is accidentally attained very early in the experimentation, it is unsure that this is the best solution beforehand. One has to continue with the GA to further generations to check if these generations produce better solutions or not. This, of course, uses a lot of needless time and resources.

3. In increasing the number of elements that are involved in the optimization process, there is an enormous increase in the search space of the

algorithm. This makes the genetic algorithm unsuitable for problems of some types or nature. In simple words, when the number of variables that are being optimized is high, the number of combinations of random mutations (changes in the levels) will increase exponentially.

4. One of the key advantages and working features of the genetic algorithm is the randomness of the process, caused by random mutations[1] Although this defines the nature of the algorithm, it is a disadvantage in some aspects. Since the algorithm uses the fact that parts of a particular solution are responsible for the poorer results, while other parts are responsible for better results. It tries to better these poorer parts, while keeping the better performing parts of the string intact. There is an issue of how to protect parts of the solution, due to the random mutations, that have evolved to represent the good solutions. Mutation to these parts may lead to loss of the progress achieved so far. Although significant research work suggests modifying the GA to help this situation, there may be a problem of not achieving a definitive generalized solution if these changes are used. Although with time and enough number of generations, one can have better solutions, it is not quite enough if the number of generations is very high making the process very time consuming. Hence, this will consume resource and time without constructive mutations.

5. One of the most frequently conceived issues with the genetic algorithm technique is the premature convergence of the algorithm.[3] It is expected that the algorithm converges at the optimal value of the fitness function. This value is in most cases the maximum value. The responses can be such that there is a local maximum in the fitness function that leads to the solution converging to this value rather that the global maximum value. Therefore, the solution obtained from the genetic algorithm happens to be a pseudo solution, which appears to vary based on the initial guess or the parent solutions (zeroth generation of solutions), and from which one starts the genetic algorithm. Thus, the genetic algorithm is unable to sacrifice short-term benefits for the sake of long-term benefits. By changing the fitness function, it is possible to resolve this issue, or by using techniques to maintain a diverse population at all times, we can manage this problem. A frequently used technique to manage the above method of maintaining diversity is to add a penalty to individuals of similar nature (similarity). This is *niche penalty*. This manages to reduce the representation of that group in the subsequent generations permitting other variants or solutions to be available in the subsequent generations. This trick, however, may not be effective, depending on the nature of the problem. Another way to deal with this issue is just to replace the part of the solution population with some randomly generated individual solutions when it appears

that most of the population is very similar to each other. This is to maintain the diversity factor of the GA as crossing over from a set of similar population. This yields to similar and repetitive solutions without any benefit. However, this nature of the genetic algorithm is an obstacle only for problems of particular nature and shape of fitness curves.

6. Genetic algorithm is not very effective for solving problems that have a yes or a no answer.[7] Using a genetic algorithm for such a problem is not good as there is no clear direction. Therefore, the evolutionary nature of the GA, which was giving us a huge edge, is no longer useful or needed. Even a random search, being simpler and computationally inexpensive, would serve the purpose better. If the problem has success or failure to be repeated giving different results, then the ratio of successes to failures provides a suitable fitness measure. GA is not the ideal way to search, as there are no meaningful fitness functions for comparison among different sets of solutions, namely, to check the ability of a particular operational point, crossing a given constraint, not for the best solution available.

7. The genetic algorithm is a very generic approach to solve a problem.[4] For a specific problem, specific search algorithms may perform better than the GA for a given amount of computation time or experimental time and resources. However, GA is a very generic algorithm that finds wider applications than the other more specific algorithms.

7.6 How GA Finds Uses in Biological Systems?

In recent years, genetic algorithm has become a very important and one of the most frequently used algorithms with respect to research conducted in the biological reactions and systems. One of the prime reasons for this is the high efficiency of this mode of optimization compared to the other techniques.[1] Although it is criticized for needing a large computational power and for being very indefinitive and vague, with a precise termination condition, it can do wonders.

In most industries, one looks for achieving the maximum production. It is often a case of slowly improving the production and hence the profit in the system. In a genetic algorithm approach for optimization, the termination condition most often achieves a particular level of productivity, as expected.[7] In using other means of optimization or experimental designs, in most cases, the design is fixed that leads one to run an innumerable number of experiments. This increase is almost exponential when the number of levels at which the reactor has to be operated is increased. This is obviously the case in a real industry as there is a continuous set or range of operating conditions available to the experimenter. To pick the best operating conditions, it would be better to discretize them with as many levels as possible.

Therefore, it is better to use genetic algorithms compared to conventional designs. GA also has the added advantage of not making the experimenter do redundant experiments. In conventional designs, it is often the case that the experimenter does many experiments. Then one has to choose the best available option. Let us assume that one does an experiment with say five variables. The experimenter figures out that variable 1 gives the best results when it is kept at a lower level just from some of the possible experiments. Still without using this acquired knowledge in conventional designs, one ends up doing all the experimental combinations, wasting time and important resources. However, unlike the case mentioned above, the genetic algorithm continues to use the previously acquired knowledge to decide where to perform the experiment next.

The most commonly used situation of genetic algorithm in biological systems is in medium optimization. In biological systems, the cells synthesize the product often react or respond to the external conditions of the environment. Both the physical and the chemical environments of the reactor have a huge effect on the productivity of the cell. Therefore, choosing the best combinations, or in other words, composition of the environment or the chemical medium is of utmost importance in biological systems.

Genetic algorithm is widely used to serve this purpose. Initially, different components that are believed to have a significant effect on the productivity are chosen – most likely they are chosen based on previous studies or experiments conducted on this system.[7] Then they are optimized in a stepwise fashion with the productivity used as the fitness function.

7.6.1 Practical Applications

As mentioned earlier, it is possible to solve mathematical optimization problems using a genetic algorithm. This makes it a very useful tool in helping one to decide the operational conditions for biological systems. When a cell culture grows in an artificial medium, a number of factors require optimization in order to get the best result or output possible. Now, the genetic algorithm finds extensive use for optimizing these factors. Thus, the major applications of genetic algorithm in biological systems include the following:[7]

1. *Temperature*: A variety of characteristics of the biological reactions like the rate[8] and feasibility depend on the temperature of operation. Apart from this, if the temperature is not within the specified limits, cells that synthesize the product will no longer be alive or the enzymes will be degraded, failing to produce even if the product is stable under the given temperature conditions.[5]

Most bacteria can grow over a temperature range of about 30°C or more, but have a narrow range for optimal growth. Temperature below and above the optimum[7] growth rate declines, defining the minimum and maximum growth temperatures, respectively.

2. *pH*: The internal pH of bacteria is fairly close to neutral. Some pH conditions maintain to maximize the output or product production of the microorganism. Most microorganisms can be classified under one of the following three categories based on their optimal pH growing conditions: acidophiles, neutrophiles, and basophiles.[7]

3. *Aeration/oxygen level*: Oxygen is of utmost importance for the growth of an aerobic cell. Some cells thrive under conditions with high amount of oxygen, while some others die under the same conditions,[1] but perform very well when there is no oxygen in the medium. In general, they are aerobic, anaerobic, facultative aerobes, aerotolerant aerobes, and microaerophiles.[7]

4. *Salinity*: The salinity basically is a measure of osmotic balance.[1] If the solution is too saline, then the water inside the cell comes out through the semi-permeable membrane of the cell through the process of osmosis and the cell shrinks to death. Similarly, if the salinity is very low, then the cell swells up due to uptake of water from the surrounding and it might lyse. Microorganisms that grow in extremely high-salt environments (15.6%–30% NaCl) are halophiles. Halotolerant grows on very low-salt environments.

5. *Water*: Microbes cannot grow in pure water.[7] The a_w or water activity denotes the available water.[1] This is the saturation percentage that has been achieved – a_w= partial pressure. The saturation pressure of pure water is 1, while that of a saturated salt solution is 0.75.[1] Most spoilage bacteria require a minimum a_w of 0.90. Some bacteria, yeast, and moulds can tolerate a value of a_w above 0.75. Most yeast requires even higher water activity.

6. *Nutrients require for growth*: Carbohydrates, fats, proteins, vitamins, minerals, and water are necessary for the growth of microorganisms.[8] Microbes differ in their abilities to use different nutrient sources. Differences in the utilization of nutrients and waste products they produce are important characteristics in differentiating between organisms.

7. *Light source and chemicals*: Ultraviolet light[7] affects the growth of organisms. Chemicals, such as hydrogen peroxide and chlorine, can kill or deactivate microbes. Some of these factors can be manipulated using genetic algorithm.

8. *Culture inoculation time*: In order to grow an organism in a reactor, the experimenter needs to transfer a small amount of inoculum. The time necessary for the preparation of inoculum and its transfer to the desired cultivation system varies with the nature of the individual organism.

9. *Culture inoculation medium*: A different and a more enriched medium is necessary during the process of inoculation. This medium has various compounds and elements, which is necessary at different levels.

10. *Type of reactor*: The type of reactor, namely, batch, continuous, and semi-batch, involves different production strategy.[9] Each reactor type is given a lower or a higher level randomly. The genetic algorithm can be applied using the above strategy, where one or more specific bits of the string represent the reactor type.

7.6.2 Solved Examples

Example 7.1

Let us consider a problem to be solved using genetic algorithm. The problem statement refers below.

For a biological reaction, conditions of operation are necessary for optimization to produce the maximum amount of cells. The variables under control of the experimenter are as follows:

1. Temperature of operation
2. pH of the reaction medium
3. The type of the reactor – bottom mounted or top mounted
4. Aeration rate
5. Culture pre-inoculation time
6. Reactor diameter – for constant volume operation

Each of the above variables can be represented in two different levels: high and low levels of the variable. Make an attempt to solve the above problem using a genetic algorithm. Indeed simpler methods like factorial design of experiments or other methods solve this trivial problem. For systems that are more complex or even for the same system, with each variable being able to take about four or five different levels, the other methods become infeasible due to their extensive nature.

Since the optimization of the amount of cells produced is the fitness function, the cell density produced after the reaction time is of consideration. Finally, a termination condition is necessary for the genetic algorithm. Both the termination conditions like the number of generations or a particular cell density are the required information.

Solution:
We follow the flowchart (*cf.* Figure 7.1) for solving optimization problems using genetic algorithm.

Steps:
Step 1 is the conversion of the given problem into binary form. This problem has six factors. Each can take two possible levels. One can use a six-digit binary string for the purpose of representation.

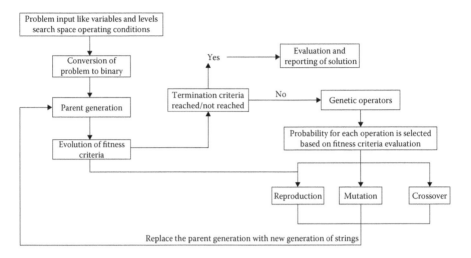

FIGURE 7.1
GA at a glance.

Let 0s represent the lower level of the variable in consideration. 1s represent the higher level of the variable. The order of the variables represented is as follows: temperature, pH, reactor type, aeration rate, culture pre-inoculation time, and reactor diameter.

For example, the desired condition is high level of temperature, lower level of pH, lower level of reactor type (any one of the reactors can be considered lower and the other higher), lower level of aeration rate, higher level of the culture pre-inoculation time, and lower level of the reactor diameter. This is given as follows:

$$100010$$

Now, the problem is converted into binary representation.
Step 2 of the problem is to generate the parent population. There are totally 2^6 possibilities. One considers five at a given time (parallelism as talked about in the explanation of genetic algorithm). These generations are random. This is done by tossing a fair coin 30 times. The following parental population is as follows:

1. 100111
2. 010011
3. 000011
4. 110111
5. 101001

In *Step 3*, evaluation of the fitness criteria is as per the flow diagram (Figure 7.1). This needs to be done experimentally. The conditions as prescribed by binary representation are decoded and experiments conducted at these mentioned conditions.

Once these experiments are carried out under the required conditions, then the results, in this case the cell density of the reactor, are to be measured and noted. Let one assume the following output or response for the system:

Case 1 – 2 units
Case 2 – 4 units
Case 3 – 1 unit
Case 4 – 1 unit
Case 5 – 2 units

After having the responses, the next step (*Step 4*) is to check for attaining termination. For the sake of the problem, the assumption for the termination criteria is an output response of 10 units. Since the termination condition is not yet satisfied, it is necessary to generate the next set of offspring or the next generation of solutions using the genetic algorithm (genetic operators).

To generate the next generation of solutions, one can make four of the next generation strings by crossover, and one of them is by a mutation to an existing string (parent solution).

During any of the selection process, the process is not completely random or uniform selection, but it is random selection weighted upon the responses. For this purpose, a probability table suggests for picking each of the string. The probability is just the response of the given parent divided by the sum of all the responses in that generation. Parents are as follows:

1. Parent 1 – 0.2
2. Parent 2 – 0.4
3. Parent 3 – 0.1
4. Parent 4 – 0.1
5. Parent 5 – 0.2

The selection of each string considers the above probabilities for crossover. MATLAB® or similar software is capable of generating random number. A random integer number is generated between 1 and 10. If it is 1 or 2, parent 1 is selected. If it is 3, 4, 5, or 6, the selection is parent 2. For 7, it is parent 3. If it is 8, parent 4 is selected. Otherwise, the selection is parent 5.

In this example, the selection of two such parents followed by the generation of another random number between 1 and 5, which gives the digit from which crossover has to be performed.

For example, 6 and 1 are the random numbers generated for the selection of the parents for crossover. Parents 1 and 2 are for crossover. The next random number for point of cross of decision turns out to be 2. This means that one crossover after the second digit. The offspring are the following:

Generation1_1 – 100011
Generation1_2 – 010111

The repetition of the above procedure gives Generation1_3 and Generation1_4.

For this case, the first two random numbers generated are 7 and 9 and the next one is 3.

Therefore,

Generation1_3 – 000001

Generation1_4 – 101011

For generation 5, one parent is selected as mentioned above. The generation of a random number is again between 1 and 6, but for the digit, the mutation is of choice. The selected digit is changed from 1 to 0 or vice versa.

The corresponding random numbers generated are 8 and 1. Therefore, the generation is Generation1_5.

Generation1_5 – 010111

From the loop again, as demonstrated by the schematic diagram (Figure 7.1), the experiments repeated at the given conditions (by gen 1 strings). One records the responses. A probability chart is created as above. On that basis, the generation 2 solutions come from using the generation 1 solutions as the parents. Repetition continues until the termination criteria. In this case, until cell density is above 10 units and then reports the solution.

Other variants of genetic algorithm/operators like different kinds of crossover and mutations can be made to make the solution more accurate and less time and resource consuming depending on the complexity of the problem. For such a simple case, it is not necessary to use such techniques.

7.7 Hybrid Design of Experiments Based on GA

7.7.1 Box-Behnken and GA-Based Hybrid Design

The Box-Behnken design is an independent quadratic design in that it does not contain an embedded factorial or fractional factorial design.[7] This is a design of experiments used to develop models for systems that are hard to operate at the extreme conditions of all or multiple variables at the same time. Limited orthogonal blocking in the Box-Behnken design compared to the central-composite designs.

For instance, the Box-Behnken design for three factors involves three blocks, in each of which two factors are varied through the four possible combinations of high and low. It is necessary to include centre points as well (in which all factors are at their central values).[7]

7.7.2 Central-Composite Design and GA-Based Hybrid Design

A central-composite design contains factorial or fractional factorial design. Estimation of curvature is due to the centre points and axial points.[7] If the distance from the centre of the design space to a factorial point is ±1 unit along the direction of each factor, the distance from the centre of the design space to a star point is $\pm\alpha$. The precise value of α depends on certain properties desired for the design and on the number of factors involved. The number of centre point runs of the design also depends on certain properties required for the design. The central-composite design of experiments gives option for more number of experiments to be performed, but it is a *very accurate way to determine higher order models.*

7.7.3 Integration with GA

There are possibly two ways of integrating the GA with these above-mentioned designs of experiments.

The first method involves the case of multiple responses. For example, an organism synthesizes two or more products. The aim is to maximize the output of the weighted sum of the products. It requires knowing beforehand what the weights for each of the individual products are.

Therefore, to solve this problem, the problem may have two parts. The first part of the problem involves solving or figuring out a response equation for the individual responses as if they are the only quantities of interest.

To do this, design of experiments approach like Box-Behnken or central-composite design is a good choice. Then mathematical models that describe the responses individually as a function of the variables considered are the final form of solution. It can also be integrated with the genetic algorithm leading to a more robust solution, the steps of which are mentioned below.

A fitness function based on the sum of the weighted values of the responses is defined for the problem. Now, a binary or a coded representation of the solutions to the particular problem is mapped out and finalized. Random selection of parent population of the variables is the next step followed by calculating the individual responses of each of the product using the mathematical models described. Next is the evaluation of fitness criteria. After evaluation of fitness criteria for each of the parent strings, the usual algorithm follows making offspring through reproduction and mutation or its variants. Thus, using a combination of genetic algorithm and conventional design of experiments, one can find the optimal point of operation for the best operation of the plant.

The second way of integrating the genetic algorithm and the other designs includes taking the base population from the other designs as the random parent population for the genetic algorithm.

The set of experimental points given by a Box-Behnken or a central-composite design of experiments is the parental set. Therefore, the randomness may be lost, but the whole aim of choosing a random parental set is that one could get an unbiased uniform initial population. If an unbiased uniform initial population is not available, it is not possible to select some strings irrespective of the population size or the generation number. Since these experimental designs as such have a uniform nature, it is well suited to be the parental population.

The need to employ genetic algorithm even after using the basic experimental design is that these only check for the output or responses in specific regions and models the responses. In the points where there have not been any real experiments, there is always a possibility of incorporating mistake.

On the other hand, a factorial design of experiment completely covers all the possible values or combination that would cover in a genetic algorithm approach. Genetic algorithm is much shorter and easier than a complete factorial design of experiments. The factorial design is something that can be used well for optimization problems where there are two or three variables involved. For cases with four or more variables, it almost becomes impossible because of the sheer number of experiments.

Therefore, a genetic algorithm coupled with the conventional design of experiments is a method of starting with the lower level approach as a base set. It will then reach the optimal response in a systematic manner where the upper limit is the rigorous factorial design of experiment. For an unlucky experimenter, the entire process of the factorial design of experiments is necessary along with all the possible experiments.

In a genetic algorithm approach, a termination criterion is generally a target response. There might be multiple points of operations that could be able to achieve this target, but it will not be able to find out all of them using GA alone. This is possible with support from factorial design of experiments. At least one of these points will be located with much lesser effort using a combined effort with GA, which will serve most industrial purposes.

7.8 Relevant Problems and Their Solution

Industrially relevant systems are the examples in this case. They are simulated and optimized using genetic algorithm approach. The results are comparable with other conventional designs.

The first problem uses GA to optimize a system. The analysis uses a factorial design of experimental approach. The next analysis is with a system that uses a Box-Behnken design of experiment. Finally, the optimization of a couple of systems combines GA with the central-composite design of experimentation.

Example 7.2: Factorial design of experiments

Solution:
For a four-variable system, the total number of experiments is 2^4 or 3^4 for two levels (-1, $+1$) or three levels (-1, 0, $+1$), respectively. With 3^4 factorial design systems, the total number of experiments is 81. This is a difficult situation for an experimenter. However, this is demonstrated here in Table 7.1.

TABLE 7.1

3^4 Factorial Design of Experiments with the Response as HMGCoA Reductase Activity during Fermentation

Experiment Number	Variable 1	Variable 2	Variable 3	Variable 4	Response (U)
1	−1	−1	−1	−1	0.123
2	−1	−1	−1	0	0.15
3	−1	−1	−1	1	0.16
4	−1	−1	0	−1	0.11
5	−1	−1	0	0	0.18
6	−1	−1	0	1	0.19
7	−1	−1	1	−1	0.14
8	−1	−1	1	0	0.2
9	−1	−1	1	1	0.35
10	−1	0	−1	−1	0.22
11	−1	0	−1	0	0.2
12	−1	0	−1	1	0.18
13	−1	0	0	−1	0.23
14	−1	0	0	0	0.25
15	−1	0	0	1	0.4
16	−1	0	1	−1	0.3
17	−1	0	1	0	0.32
18	−1	0	1	1	0.4
19	−1	1	−1	−1	0.16
20	−1	1	−1	0	0.17
21	−1	1	−1	1	0.21
22	−1	1	0	−1	0.25
23	−1	1	0	0	0.19
24	−1	1	0	1	0.35
25	−1	1	1	−1	0.29
26	−1	1	1	0	0.41
27	−1	1	1	1	0.45
28	0	−1	−1	−1	0.12
29	0	−1	−1	0	0.15
30	0	−1	−1	1	0.16
31	0	−1	0	−1	0.15
32	0	−1	0	0	0.2
33	0	−1	0	1	0.23
34	0	−1	1	−1	0.18
35	0	−1	1	0	0.25
36	0	−1	1	1	0.33
37	0	0	−1	−1	0.16
38	0	0	−1	0	0.18
39	0	0	−1	1	0.2
40	0	0	0	−1	0.2

(Continued)

TABLE 7.1 (*Continued*)

3^4 Factorial Design of Experiments with the Response as HMGCoA Reductase Activity during Fermentation

Experiment Number	Variable 1	Variable 2	Variable 3	Variable 4	Response (U)
41	0	0	0	0	0.31
42	0	0	0	1	0.25
43	0	0	1	−1	0.18
44	0	0	1	0	0.22
45	0	0	1	1	0.4
46	0	1	−1	−1	0.17
47	0	1	−1	0	0.15
48	0	1	−1	1	0.15
49	0	1	0	−1	0.19
50	0	1	0	0	0.21
51	0	1	0	1	0.25
52	0	1	1	−1	0.14
53	0	1	1	0	0.23
54	0	1	1	1	0.36
55	1	−1	−1	−1	0.12
56	1	−1	−1	0	0.1
57	1	−1	−1	1	0.19
58	1	−1	0	−1	0.15
59	1	−1	0	0	0.24
60	1	−1	0	1	0.2
61	1	−1	1	−1	0.21
62	1	−1	1	0	0.25
63	1	−1	1	1	0.28
64	1	0	−1	−1	0.16
65	1	0	−1	0	0.15
66	1	0	−1	1	0.16
67	1	0	0	−1	0.17
68	1	0	0	0	0.3
69	1	0	0	1	0.38
70	1	0	1	−1	0.26
71	1	0	1	0	0.23
72	1	0	1	1	0.4
73	1	1	−1	−1	0.16
74	1	1	−1	0	0.15
75	1	1	−1	1	0.18
76	1	1	0	−1	0.14
77	1	1	0	0	0.14
78	1	1	0	1	0.23
79	1	1	1	−1	0.19
80	1	1	1	0	0.3
81	1	1	1	1	0.45

For optimization of this problem, a polynomial fit is made to this data set. The system is modelled algebraically. A matrix approach applies in this case.

$$AX = Y \qquad (7.1)$$

So

$$A = YX^{-1}$$

where:
X^{-1} is the inverse of the matrix X

This is the correct approach for a system where the coefficients are to be solved for an exactly required number of points from experimentation. Here, the number of points from experimentation is much higher than the number of points required to determine the coefficients or in other word the matrix A.

The aim is not to solve for an exact matrix A, but to perform regression and fit a curve that best describes the given system. Since this system has been modelled by using the factorial design of experimentation, a quadratic model fit applies best to the system to get a solution for A.

Multiplying both sides of Equation 7.1 by X' transpose of X gives Equation 7.2.

$$A(XX')^{-1} = YX' \qquad (7.2)$$

Now, multiplying both sides by $(XX')^{-1}$ gives $AI = YX'(XX')$, where I is the identity matrix.

Matrix fitting
   A MATLAB code is used for achieving the above-mentioned regression.
   Table MATLAB code-1.

```
X = [1  -1 -1 -1 -1  1  1  1  1  1  1  1  1  1  1;
     1  -1 -1 -1  0  1  1  0  1  0  0  1  1  1  0;
     1  -1 -1 -1  1  1  1 -1  1 -1 -1  1  1  1  1;
     1  -1 -1  0 -1  1  0  1  0  1  0  1  1  0  1;
     1  -1 -1  0  0  1  0  0  0  0  0  1  1  0  0;
     1  -1 -1  0  1  1  0 -1  0 -1  0  1  1  0  1;
     1  -1 -1  1 -1  1 -1  1 -1  1 -1  1  1  1  1;
     1  -1 -1  1  0  1 -1  0 -1  0  0  1  1  1  0;
     1  -1 -1  1  1  1 -1 -1 -1 -1  1  1  1  1  1;
     1  -1  0 -1 -1  0  1  1  0  0  1  1  0  1  1;
     1  -1  0 -1  0  0  1  0  0  0  0  1  0  1  0;
     1  -1  0 -1  1  0  1 -1  0  0 -1  1  0  1  1;
     1  -1  0  0 -1  0  0  1  0  0  0  1  0  0  1;
     1  -1  0  0  0  0  0  0  0  0  0  1  0  0  0;
     1  -1  0  0  1  0  0 -1  0  0  0  1  0  0  1;
     1  -1  0  1 -1  0 -1  1  0  0 -1  1  0  1  1;
     1  -1  0  1  0  0 -1  0  0  0  0  1  0  1  0;
     1  -1  0  1  1  0 -1 -1  0  0  1  1  0  1  1;
```

```
1 -1  1 -1 -1 -1  1  1 -1 -1  1  1  1  1  1;
1 -1  1 -1  0 -1  1  0 -1  0  0  1  1  1  0;
1 -1  1 -1  1 -1  1 -1 -1  1 -1  1  1  1  1;
1 -1  1  0 -1 -1  0  1  0 -1  0  1  1  0  1;
1 -1  1  0  0 -1  0  0  0  0  0  1  1  0  0;
1 -1  1  0  1 -1  0 -1  0  1  0  1  1  0  1;
1 -1  1  1 -1 -1 -1  1  1 -1 -1  1  1  1  1;
1 -1  1  1  0 -1 -1  0  1  0  0  1  1  1  0;
1 -1  1  1  1 -1 -1 -1  1  1  1  1  1  1  1;
1  0 -1 -1 -1  0  0  0  1  1  1  0  1  1  1;
1  0 -1 -1  0  0  0  0  1  0  0  0  1  1  0;
1  0 -1 -1  1  0  0  0  1 -1 -1  0  1  1  1;
1  0 -1  0 -1  0  0  0  0  1  0  0  1  0  1;
1  0 -1  0  0  0  0  0  0  0  0  0  1  0  0;
1  0 -1  0  1  0  0  0  0 -1  0  0  1  0  1;
1  0 -1  1 -1  0  0  0 -1  1 -1  0  1  1  1;
1  0 -1  1  0  0  0  0 -1  0  0  0  1  1  0;
1  0 -1  1  1  0  0  0 -1 -1  1  0  1  1  1;
1  0  0 -1 -1  0  0  0  0  0  1  0  0  1  1;
1  0  0 -1  0  0  0  0  0  0  0  0  0  1  0;
1  0  0 -1  1  0  0  0  0  0 -1  0  0  1  1;
1  0  0  0 -1  0  0  0  0  0  0  0  0  0  1;
1  0  0  0  0  0  0  0  0  0  0  0  0  0  0;
1  0  0  0  1  0  0  0  0  0  0  0  0  0  1;
1  0  0  1 -1  0  0  0  0  0 -1  0  0  1  1;
1  0  0  1  0  0  0  0  0  0  0  0  0  1  0;
1  0  0  1  1  0  0  0  0  0  1  0  0  1  1;
1  0  1 -1 -1  0  0  0 -1 -1  1  0  1  1  1;
1  0  1 -1  0  0  0  0 -1  0  0  0  1  1  0;
1  0  1 -1  1  0  0  0 -1  1 -1  0  1  1  1;
1  0  1  0 -1  0  0  0  0 -1  0  0  1  0  1;
1  0  1  0  0  0  0  0  0  0  0  0  1  0  0;
1  0  1  0  1  0  0  0  0  1  0  0  1  0  1;
1  0  1  1 -1  0  0  0  1 -1 -1  0  1  1  1;
1  0  1  1  0  0  0  0  1  0  0  0  1  1  0;
1  0  1  1  1  0  0  0  1  1  1  0  1  1  1;
1  1 -1 -1 -1 -1 -1 -1  1  1  1  1  1  1  1;
1  1 -1 -1  0 -1 -1  0  1  0  0  1  1  1  0;
1  1 -1 -1  1 -1 -1  1  1 -1 -1  1  1  1  1;
1  1 -1  0 -1 -1  0 -1  0  1  0  1  1  0  1;
1  1 -1  0  0 -1  0  0  0  0  0  1  1  0  0;
1  1 -1  0  1 -1  0  1  0 -1  0  1  1  0  1;
1  1 -1  1 -1 -1  1 -1 -1  1 -1  1  1  1  1;
1  1 -1  1  0 -1  1  0 -1  0  0  1  1  1  0;
1  1 -1  1  1 -1  1  1 -1 -1  1  1  1  1  1;
1  1  0 -1 -1  0 -1 -1  0  0  1  1  0  1  1;
1  1  0 -1  0  0 -1  0  0  0  0  1  0  1  0;
1  1  0 -1  1  0 -1  1  0  0 -1  1  0  1  1;
1  1  0  0 -1  0  0 -1  0  0  0  1  0  0  1;
1  1  0  0  0  0  0  0  0  0  0  1  0  0  0;
1  1  0  0  1  0  0  1  0  0  0  1  0  0  1;
1  1  0  1 -1  0  1 -1  0  0 -1  1  0  1  1;
1  1  0  1  0  0  1  0  0  0  0  1  0  1  0;
1  1  0  1  1  0  1  1  0  0  1  1  0  1  1;
1  1  1 -1 -1  1 -1 -1 -1 -1  1  1  1  1  1;
1  1  1 -1  0  1 -1  0 -1  0  0  1  1  1  0;
```

```
1   1  1 -1  1  1 -1  1 -1  1 -1  1  1  1  1;
1   1  1  0 -1  1  0 -1  0 -1  0  1  1  0  1;
1   1  1  0  0  1  0  0  0  0  0  1  1  0  0;
1   1  1  0  1  1  0  1  0  1  0  1  1  0  1;
1   1  1  1 -1  1  1 -1  1 -1 -1  1  1  1  1;
1   1  1  1  0  1  1  0  1  0  0  1  1  1  0;
1   1  1  1  1  1  1  1  1  1  1  1  1  1  1;]
```

% We write the design matrix of the factorial design for 4
variables
% system
% 1st column is the constant "1" and 2nd, 3rd, 4th and 5th columns
% represent the variables in their coded forms while columns
6, 7, 8, 9, 10, 11
% represent the product of variables with one another
sequentially
% The columns 12, 13,14 and 15 show the squared values of
each of the
% individual variables

```
Y = [0.123
     0.15
     0.16
     0.11
     0.18
     0.19
     0.14
     0.2
     0.35
     0.22
     0.2
     0.18
     0.23
     0.25
     0.4
     0.3
     0.32
     0.4
     0.16
     0.17
     0.21
     0.25
     0.19
     0.35
     0.29
     0.41
     0.45
     0.12
     0.15
     0.16
     0.15
     0.2
     0.23
     0.18
     0.25
     0.33
```

```
        0.16
        0.18
        0.2
        0.2
        0.31
        0.25
        0.18
        0.22
        0.4
        0.17
        0.15
        0.15
        0.19
        0.21
        0.25
        0.14
        0.23
        0.36
        0.12
        0.1
        0.19
        0.15
        0.24
        0.2
        0.21
        0.25
        0.28
        0.16
        0.15
        0.16
        0.17
        0.3
        0.38
        0.26
        0.23
        0.4
        0.16
        0.15
        0.18
        0.14
        0.14
        0.23
        0.19
        0.3
        0.44]


% The Y matrix gives the list of responses got through
experimentation at
% the specified conditions correspondingly
A = X\Y
% This command makes us fit the best possible set of values
for the matrix
% A which represents the coefficients that determine the
quadratic fit we
% intended
```

Results of model fitting

Now having run the code, we get the coefficients. The system represents the following model using these coefficients.

$$z = 0.237 - 0.0128x_1 + 0.0214x_2 + 0.0618x_3 + 0.0485x_4 - 0.0188x_1x_2 - 0.0024x_1x_3$$

$$+ 0.0012x_1x_4 + 0.0112x_2x_3 + 0.0043x_2x_4 + 0.037x_3x_4$$

$$+ 0.0191x_1^2 - 0.0414x_2^2 - 0.0014x_3^2 + 0.013x_4^2$$

where:

x_1 represents variable 1
x_2 represents variable 2
x_3 represents variable 3
x_4 represents variable 4

The system is successfully modelled using factorial design of experiments.

Absolute maxima

Now, the above equation is differentiated to evaluate the maxima or to decide the maxima of the function. Result is the four differential equations that have to be solved for obtaining maxima. Then, they are solved simultaneously to yield the point at which this function achieves its maximum value. It could also be the minimum value, but this can be checked by simply comparing that value with the value of the function at some other point. On the other hand, we can use the genetic algorithm to help us in this process by using the above-developed equation as the fitness function for our GA. Now we optimize to get results as mentioned below.

This yields the point [8.33 2.77 18.93 21.96], at a fitness value of 24.70 after optimization using genetic algorithm for 100 generations at the parameter values used below for other simulations.

This is not a practically feasible point. The optimum achieved is not something that can be used in reality as the values of the variables lie well outside that of the operable range of the variables, which is [−1 1].

Maxima within operable range

The goal of the problem is to optimize this system within the experimental limits. For this, GA couples with factorial design approach and comes up with the optimal point of operation.

The above-mentioned model is used as the fitness function for the GA for optimization of the problem under the constraints of the values of the variables being within −1 and 1. The genetic algorithm optimization is done using the GA toolbox of MATLAB.

The GA toolbox of MATLAB is programmed to minimize a function and not to maximize it, so while writing the fitness function, it is written as the negative of the actual fitness function and minimized, thereby leading to maximization of the actual fitness function.

This optimization yields a fitness function value of 0.44204

The parameter values are as follows:

Number of generations = 52
Population size = 20 individuals

Scattered or random type crossover
Elite count used was 2
Crossover fraction used 0.8
The simulation takes 3.2 s

These parameter values were chosen after multiple trials and led one to believe that these values would most likely converge on the best value of the fitness function in the most robust manner. The fitness function used was a negative of the z-function (response function). The stopping criterion used was either 100 generations or precision of the GA increment falling below that of a given value.

The value of the variables at which this value was achieved was given by [−0.99 0.67 0.99 0.99]. This gives the optimization using GA.

Optimization on integer points
Since the domain of the design comprises the point obtained, there is no problem in operating at these conditions. In practice, for some biological systems, it might not even be feasible to run the plant at a particular condition even if it is possible to run it at conditions either side of it.

As the next step, optimization using GA with a constraint studied at the values of the parameters only for integer values (−1,0,1,1). This becomes a very realistic case, as one knows that the system may definitely be operated at these conditions.

A simulation with the above-mentioned integer constraint gives the fitness function. Hence, fitness function value = 0.438.

Values of the parameters are as follows:

Number of generations = 48
Population size = 20 individuals

Crossover type used is scattered or random type
Elite count used = 2
Crossover fraction used = 0.8
The simulation takes 2.4 s

The optimum is attained at [-1 1 1 1].

This point compares well with that obtained from experimentation using the factorial design of experiments. The optimum point from experimentation coincides with the simulated values. The response got from the optimization is 0.43, while the value from experimentation is 0.45. This difference is not due to the factorial design, but it is due to imperfections during the modelling performed by the quadratic polynomial using the factorial design.

Example 7.3: Comparison of GA and factorial design

To compare the performance of the factorial design of experiments with GA, optimization analysis for Example 7.2 has been done separately.

Method:
Experiments have been performed following the factorial design approach. Then the system is modelled using quadratic equations as per earlier procedure. To predict optima, the model links with some optimization technique, which will again use some computational power.

Now, the genetic algorithm, in this case, cannot be performed computationally as that requires a continuous fitness function. One needs to discretize the values. The next step will only be known after performing an experiment, which will give a fitness value that leads to the next step. It is to be noted that a GA is not a reproducible algorithm. Thus, the experimenter needs to run the algorithm multiple times before one makes a generalized conclusion regarding the comparison of the GA and the factorial design. One can also hope to draw conclusion about the extreme cases namely, where the GA needed the maximum and the minimum amount of experiments to be performed.
We perform the GA by hand under the following conditions:

Fitness function is the response obtained from experimentation.
Stopping criterion is the fitness value of 0.40. This is chosen after looking at the industrial yield currently and the improvement to be achieved.
Number of generations is not constrained, but it will use the above stopping criterion or else stop when 50 generations are reached.
Crossover type is random or scattered.
Elite fraction is 2
Mutation probability $= 0.1$
Crossover fraction used $= 0.8$
Population size $= 10$

One uses a random number generator from MATLAB to maintain the randomness in decision making in order to achieve our results.
The results obtained are

Average final fitness value $= 0.41$
Maximum final fitness achieved $= 0.45$
Average number of generations $= 7$
Average distinct points used $= 26.7 = 27$
Average number of experiments performed $= 27$
Maximum number of experiments to be performed in any one algorithm trial $= 49$

Result
GA does a much better job than the factorial design of experimentation. The factorial design requires 81 experiments, whereas GA requires only 27 (average). The only advantage of the factorial design is to operate the system at fractional or intermediate values, where the fitted model comes in handy. This is not such a big factor, especially in biological systems, where immense importance is given to the precision of conditions at which the system is operated. Therefore, if one operates in such random conditions, the system may be faulty and even slight deviations may

prove risk to the system. The model fitted by the factorial design is not very accurate. There is no guarantee that it will work well for points outside the domain of the factorial design like fractional points.

Thus, clearly the GA performs better than factorial design for determining the optimal value of the responses of biological systems.

Example 7.4: Central-composite design system

Table 7.2 shows the data of the experimental plan. The value of α is calculated for an orthogonal design as 2. This is used when we are designing the experimental points as shown below. In addition, six centre point trials are used.

In order to optimize this problem, a polynomial is fitted to this data set. This is done using a matrix approach as follows (Equation 7.1):

TABLE 7.2

Central-Composite Orthogonal Design Matrix with the Corresponding Response

Experiment Number	Variable 1	Variable 2	Variable 3	Variable 4	Response
1	−1	−1	−1	−1	41.44
2	−1	−1	−1	1	65.14
3	−1	−1	1	−1	37.13
4	−1	−1	1	1	56.24
5	−1	1	−1	−1	41.84
6	−1	1	−1	1	63.24
7	−1	1	1	−1	37.85
8	−1	1	1	1	56.61
9	1	−1	−1	−1	40.55
10	1	−1	−1	1	65.11
11	1	−1	1	−1	38.66
12	1	−1	1	1	56.83
13	1	1	−1	−1	41.59
14	1	1	−1	1	63.67
15	1	1	1	−1	38.27
16	1	1	1	1	59.56
17	2	0	0	0	45.91
18	−2	0	0	0	46.37
19	0	−2	0	0	39.52
20	0	2	0	0	48.28
21	0	0	−2	0	48.75
22	0	0	2	0	40.55
23	0	0	0	−2	48.2
24	0	0	0	2	60.51
25	0	0	0	0	49.28
26	0	0	0	0	48.58
27	0	0	0	0	50.12
28	0	0	0	0	50.68
29	0	0	0	0	50.47
30	0	0	0	0	48.67

$$AX = Y \tag{7.1}$$

Thus yielding $A = YX'\left(XX'\right)$ as referred in Example 7.2.

Fitting the matrix
The MATLAB code applies for achieving the regression mentioned above.
Table MATLAB code-2.

```
X =  [1 -1 -1 -1 -1   1   1   1   1   1   1   1   1   1   1;
      1 -1 -1 -1   1   1   1 -1   1 -1 -1   1   1   1   1;
      1 -1 -1   1 -1   1 -1   1 -1   1 -1   1   1   1   1;
      1 -1 -1   1   1   1 -1 -1 -1 -1   1   1   1   1   1;
      1 -1   1 -1 -1 -1   1   1 -1 -1   1   1   1   1   1;
      1 -1   1 -1   1 -1   1 -1 -1   1 -1   1   1   1   1;
      1 -1   1   1 -1 -1 -1   1   1 -1 -1   1   1   1   1;
      1 -1   1   1   1 -1 -1 -1   1   1   1   1   1   1   1;
      1   1 -1 -1 -1 -1 -1 -1   1   1   1   1   1   1   1;
      1   1 -1 -1   1 -1 -1   1   1 -1 -1   1   1   1   1;
      1   1 -1   1 -1 -1   1 -1 -1   1 -1   1   1   1   1;
      1   1 -1   1   1 -1   1   1 -1 -1   1   1   1   1   1;
      1   1   1 -1 -1   1 -1 -1 -1 -1   1   1   1   1   1;
      1   1   1 -1   1   1 -1   1 -1 -1   1   1   1   1   1;
      1   1   1   1 -1   1   1 -1   1 -1 -1   1   1   1   1;
      1   1   1   1   1   1   1   1   1   1   1   1   1   1;
      1   2   0   0   0   0   0   0   0   0   4   0   0   0;
      1  -2   0   0   0   0   0   0   0   0   4   0   0   0;
      1   0  -2   0   0   0   0   0   0   0   0   4   0   0;
      1   0   2   0   0   0   0   0   0   0   0   4   0   0;
      1   0   0  -2   0   0   0   0   0   0   0   0   4   0;
      1   0   0   2   0   0   0   0   0   0   0   0   4   0;
      1   0   0   0  -2   0   0   0   0   0   0   0   0   4;
      1   0   0   0   2   0   0   0   0   0   0   0   0   4;
      1   0   0   0   0   0   0   0   0   0   0   0   0   0;
      1   0   0   0   0   0   0   0   0   0   0   0   0   0;
      1   0   0   0   0   0   0   0   0   0   0   0   0   0;
      1   0   0   0   0   0   0   0   0   0   0   0   0   0;
      1   0   0   0   0   0   0   0   0   0   0   0   0   0;
      1   0   0   0   0   0   0   0   0   0   0   0   0   0;]

% We write the design matrix of the CCD design for 4 variables
% system

% 1st column is the constant "1" and 2nd, 3rd, 4th and 5th
columns
% represent the variables in their coded forms while columns
6,7,8,9,10,11
% represent the product of variables with one another
sequentially

%The columns12, 13, 14 and 15 show the squared values of each of the
% individual variables

Y =  [41.44
      65.14
      37.13
```

```
56.24
41.84
63.24
37.85
56.61
40.55
65.11
38.66
56.83
41.59
63.67
38.27
59.56
45.91
46.37
39.52
48.28
48.75
40.55
48.2
60.51
49.28
48.58
50.12
50.68
50.47
48.67]
```

```
% The Y matrix gives the list of responses got through
experimentation at
% the specified conditions correspondingly
```

```
A = X\Y
% this command makes us fit the best possible set of values
for the matrix
% A which represents the coefficients that determine the
quadratic fit we
% intended
```

Results

Coefficients are obtained after running the code. The following model represents the system.

$$z = 49.6333 + 0.1846x_1 + 1.0118x_2 - 2.6846x_3 + 9.1788x_4 - 0.1094x_1x_2$$

$$+0.3519x_1x_3 + 0.1331x_1x_4 - 0.0044x_2x_3 + 0.2119x_2x_4 - 1.3131x_3x_4$$

$$+0.5176x_1^2 - 1.0776x_2^2 - 0.89x_3^2 + 2.7861x_4^2$$

where:
  x_1 represents variable 1
  x_2 represents variable 2
  x_3 represents variable 3
  x_4 represents variable 4

Evaluation of maxima

Now, this equation is differentiated and the maximum is evaluated to decide the maxima of the function. Four differential equations are written by differentiating the equation with respect to each of the variables. Then they are simultaneously solved to yield the point at which this function achieves its maximum value. It could also be the minimum value, but this can be checked by simply comparing that value with the value of the function at some other point.

This yields the point [−0.724 4.73 −8.69 32.56], at a fitness value of 3644.53 after optimization by genetic algorithm for 100 generations using parameter values mentioned below for other simulations.

This is not a practically feasible point. The optimum is not something that can be used in reality as the values of the variables lie well outside that of the operable range of the variables which is [−1 1].

Optimization with experimental limits

The goal is to optimize this system within the experimental limits. For this, a GA couples with central cubic orthogonal design approach and comes up with the optimal point of operation.

The above-mentioned model acts as the fitness function for the GA, which will be optimized under the constraints of the values of the variables within −1 and 1. The genetic algorithm optimization is done using the GA tool box of MATLAB.

GA is naturally programmed to minimize a function and not maximize it. While writing the fitness function in a proper edited form such that it is written as the negative of the actual fitness function, so that the actual fitness function is maximized while this negative function is minimized.

This optimization gives a fitness function value of 65.02

The parameter values are as follows:

Number of generations = 69
Population size = 20 individuals
Scattered or random type crossover
Elite count used is 2
Crossover fraction = 0.8
The simulation takes 2.5 s

These parameter values were chosen after multiple trials that lead one to believe that these values would most likely converge on the best value of the fitness function. The fitness function used was a negative of the z-function mentioned before. The stopping criterion used was either 100 generations or precision of the GA increment falling below that of a given value.

The value of the variables at which this value was achieved is [−0.091 0.572−0.994 0.997]. Thus, the optimization is by using GA.

Now, if one looks at the results obtained, since the domain of the design comprises the point obtained, one should not have any problem in operating at these conditions. In practice, for some biological systems, it might not even be feasible to run the plant at a particular condition even if it is possible to run it at those conditions either side of it.

Integer constraint application

In the next step, optimization using GA with a constraint is applied to the system, the values of the parameters can only be integer values (−1,0,1). This becomes a realistic case since the system can definitely be operated at these conditions.

Running the simulations with the above-mentioned integer constraint, one can have a fitness function value of 64.85

The parameter values are as follows:

Number of generations = 51
Population size = 20 individuals
Scattered or random type crossover
Elite count used is 2
Crossover fraction used is 0.8
The simulation takes 1.86 s

The point at which this optima is reached is [0 1 −1 1].

Example 7.5: Central-composite design system – rotatable

Solution:
Table 7.3 is the central-composite design for four variables for a rotatable case. $\alpha = (nc)^{1/4} = 2$, assuming n_0 randomly as 2 (centre points). The grid is mentioned here. One can follow the procedure described in Examples 7.2 and 7.4.

In order to optimize this problem, one needs to fit a polynomial to this set of data and model the system algebraically. This is done using a matrix approach described in Example 7.2 and the Equation 7.1.

TABLE 7.3

Rotatable Central-Composite Design Matrix

Experiment Number	Variable 1	Variable 2	Variable 3	Variable 4	Response
1	−1	−1	−1	−1	0.0882
2	−1	−1	−1	1	0.0672
3	−1	−1	1	−1	0.0562
4	−1	−1	1	1	0.0325
5	−1	1	−1	−1	0.0904
6	−1	1	−1	1	0.0614
7	−1	1	1	−1	0.0946
8	−1	1	1	1	0.0398
9	1	−1	−1	−1	0.0078
10	1	−1	−1	1	0.0806
11	1	−1	1	−1	0.0674
12	1	−1	1	1	0.0576
13	1	1	−1	−1	0.0878
14	1	1	−1	1	0.0744

(Continued)

TABLE 7.3 (*Continued*)

Rotatable Central-Composite Design Matrix

Experiment Number	Variable 1	Variable 2	Variable 3	Variable 4	Response
15	1	1	1	−1	0.0552
16	1	1	1	1	0.0404
17	2	0	0	0	0.0124
18	−2	0	0	0	0.0502
19	0	−2	0	0	0.0356
20	0	2	0	0	0.0466
21	0	0	−2	0	0.0488
22	0	0	2	0	0.0402
23	0	0	0	−2	0.0568
24	0	0	0	2	0.0274
25	0	0	0	0	0.0614
26	0	0	0	0	0.0404

This is the correct approach for a system where the coefficients are to be solved for from an exactly required number of points from experimentation. In this case, the number of points from experimentation is much higher than the number of points required to determine the coefficients or in other word the matrix A.

The aim is not to solve for an exact matrix A but to perform regression and fit a curve that best describes the given system. Since this system has been modelled by using central cubic rotatable design of experimentation, a quadratic model fits to the system and solve for A. We do the following algebraic operation to fit the polynomial. One has to write a MATLAB code for achieving the above-mentioned regression.

Table MATLAB code-3.

```
X = [1 -1 -1 -1 -1  1  1  1  1  1  1  1  1  1  1;
     1 -1 -1 -1  1  1  1 -1  1 -1 -1  1  1  1  1;
     1 -1 -1  1 -1  1 -1  1 -1  1 -1  1  1  1  1;
     1 -1 -1  1  1  1 -1 -1 -1 -1  1  1  1  1  1;
     1 -1  1 -1 -1 -1  1  1 -1 -1  1  1  1  1  1;
     1 -1  1 -1  1 -1  1 -1 -1  1 -1  1  1  1  1;
     1 -1  1  1 -1 -1 -1  1  1 -1 -1  1  1  1  1;
     1 -1  1  1  1 -1 -1 -1  1  1  1  1  1  1  1;
     1  1 -1 -1 -1 -1 -1 -1  1  1  1  1  1  1  1;
     1  1 -1 -1  1 -1 -1  1  1 -1 -1  1  1  1  1;
     1  1 -1  1 -1 -1  1 -1 -1  1 -1  1  1  1  1;
     1  1 -1  1  1 -1  1  1 -1 -1  1  1  1  1  1;
     1  1  1 -1 -1  1 -1 -1 -1 -1  1  1  1  1  1;
     1  1  1 -1  1  1 -1  1 -1 -1  1  1  1  1  1;
     1  1  1  1 -1  1  1 -1  1 -1 -1  1  1  1  1;
     1  1  1  1  1  1  1  1  1  1  1  1  1  1  1;
     1  2  0  0  0  0  0  0  0  0  4  0  0  0;
     1 -2  0  0  0  0  0  0  0  0  4  0  0  0;
     1  0 -2  0  0  0  0  0  0  0  0  4  0  0;
     1  0  2  0  0  0  0  0  0  0  0  4  0  0;
```

```
1   0   0  -2   0   0   0   0   0   0   0   0   0   4   0;
1   0   0   2   0   0   0   0   0   0   0   0   0   4   0;
1   0   0   0  -2   0   0   0   0   0   0   0   0   0   4;
1   0   0   0   2   0   0   0   0   0   0   0   0   0   4;
1   0   0   0   0   0   0   0   0   0   0   0   0   0   0;
1   0   0   0   0   0   0   0   0   0   0   0   0   0   0;]
```

% We write the design matrix of the Box and Behnken design
For our 4 variable
% system

% 1st column is the constant "1" and 2nd, 3rd, 4th and 5th columns
% represent the variables in their coded forms while columns
6,7,8,9,10,11
% represent the product of variables with one another
sequentially

% The columns 12 , 13 ,14 and 15 show the squared values of
each of the
% individual variables

```
Y = [0.0882
     0.0672
     0.0562
     0.0325
     0.0994
     0.0614
     0.0946
     0.0398
     0.0917
     0.0806
     0.0674
     0.0576
     0.0878
     0.0744
     0.0552
     0.0404
     0.0124
     0.0502
     0.0356
     0.0466
     0.0488
     0.0402
     0.0568
     0.0274
     0.0614
     0.0404]
```

% The Y matrix gives the list of responses got through
experimentation at
% the specified conditions correspondingly

A = X\Y

% This command makes us fit the best possible set of values
for the matrix
% A which represents the coefficients that determine the
quadratic fit we
% intended

Coefficients are obtained after running the code. The following model represents the system.

$$z = 0.0509 + 0.0011x_1 - 0.0005x_2 + 0.0129x_3 - 0.0097x_4 - 0.0024x_1x_2$$

$$-0.0005x_1x_3 + 0.0002x_1x_4 - 0.002x_2x_3 - 0.006x_2x_4 + 0.0041x_3x_4$$

$$-0.0003x_1^2 + 0.022x_2^2 - 0.003x_3^2 + 0.0024x_4^2$$

where:
  x_1 represents variable 1
  x_2 represents variable 2
  x_3 represents variable 3
  x_4 represents variable 4

Thus, the system using the central cubic rotatable design of experimental approach has been successfully modelled.

Now, this equation is differentiated and the maximum is evaluated to decide the maxima of the function. Four differential equations are written by differentiating the equation with respect to each of the variables. Then they are simultaneously solved to yield the point at which this function achieves its maximum value. It could also be the minimum value, but this can be checked by simply comparing that value with the value of the function at some other point.

This yields the point, for example, [11.53 −24.76 5.4 9.01], at a fitness value of 1.609 after 100 generations of optimization using GA at the parameter value specified below.

But this is not a practically feasible point, the optimum got is not something that can be used in reality as the values of the variables lie well outside that of the operable range of the variables, which is [−1 1].

The goal is to optimize this system within the experimental limits. A GA coupled with the central cubic rotatable design approach applies to come up with the optimal point of operation.

We use the above-mentioned model as the fitness function for the GA and optimize it under the constraints of the values of the variables being within −1 and 1. The genetic algorithm optimization is done using the GA toolbox of MATLAB.

The GA is naturally programmed to minimize a function and not maximize it. While writing the fitness function, edit it such that it is written as the negative of the actual fitness function.

For example, this optimization yields a fitness function value of 0.097. The parameter values are as follows:

Number of generations = 42
Population size = 20 individuals
Crossover type used was scattered or random type
Elite count used is 2
Crossover fraction used is 0.8
The simulation takes 1.7 s

These parameter values are chosen after multiple trials that led one to believe that these values would most likely converge on the best value of the fitness function. The fitness function used a negative of the z-function mentioned before. The stopping criterion used was either 100 generations or precision of the GA increment falling below that of a given value.

The value of the variables at which this value was achieved could be given by [−0.54 0.99 0.99 −1].

If one looks at the results obtained, since the domain of the design comprises the point obtained we should not have any problem in operating at these conditions. In practice, for some biological systems, it might not even be feasible to run the plant at a particular condition, even if it is possible to run it at conditions either side of it.

In the next step, one performs optimization using GA with a constraint that the values of the parameters can only be integer values (−1,0,1). This becomes a very realistic case since one knows that the system can definitely be operated at these conditions.

For example, if the simulations run with the above-mentioned integer constraint, probable results could be the following:

Fitness function value = 0.096

The parameter values are as follows:

Number of generations = 35
Population size = 20 individuals
Scattered or random type crossover
Elite count used is 2
Crossover fraction used is 0.8
The simulation takes 1 s

The point at which this optima is reached is [−1 1 1 −1].

This point compares with that obtained from experimentation using the central cubic rotatable design of experiments. It is seen that even from experimentation the point that has the maximum optima is this point. The response got from the optimization is, for example, 64.85, while that got from the experimentation is 66.67. This difference is not due to the central cubic rotatable design, but it is probably due to the imperfect modelling using the quadratic polynomial.

Example 7.6: System modelled using the Box-Behnken design

Table 7.4 gives the Box-Behnken design of experimentation matrix with the response.

Equation 7.1 represents necessary matrix.

$$AX = Y \qquad (7.1)$$

The aim is not to solve for an exact matrix A, but to perform regression and fit a curve that best describes the given system. Since this system has been modelled by using the Box-Behnken design of experimentation, a quadratic model fits to the system for solving for A. A MATLAB code is used for achieving the above-mentioned regression.

TABLE 7.4

The Box-Behnken Design Matrix with Response

Experiment Number	Variable 1	Variable 2	Variable 3	Variable 4	Response
1	0	0	−1	−1	0.205
2	0	0	−1	1	0.405
3	0	0	1	−1	0.19
4	0	0	1	1	0.28
5	0	−1	0	−1	0.26
6	0	−1	0	1	0.38
7	0	1	0	−1	0.315
8	0	1	0	1	0.36
9	0	−1	−1	0	0.17
10	0	−1	1	0	0.438
11	0	1	−1	0	0.24
12	0	1	1	0	0.37
13	−1	0	0	−1	0.257
14	−1	0	0	1	0.36
15	1	0	0	−1	0.2
16	1	0	0	1	0.35
17	−1	0	−1	0	0.23
18	−1	0	1	0	0.482
19	1	0	−1	0	0.417
20	1	0	1	0	0.38
21	−1	−1	0	0	0.15
22	−1	1	0	0	0.34
23	1	−1	0	0	0.42
24	1	1	0	0	0.48

Table MATLAB code-4.

```
X = [1   0   0  -1  -1   0   0   0   0   0   1   0   0   1   1;
     1   0   0  -1   1   0   0   0   0   0  -1   0   0   1   1;
     1   0   0   1  -1   0   0   0   0   0  -1   0   0   1   1;
     1   0   0   1   1   0   0   0   0   0   1   0   0   1   1;
     1   0  -1   0  -1   0   0   0   0   1   0   0   1   0   1;
     1   0  -1   0   1   0   0   0   0  -1   0   0   1   0   1;
     1   0   1   0  -1   0   0   0   0  -1   0   0   1   0   1;
     1   0   1   0   1   0   0   0   0   1   0   0   1   0   1;
     1   0  -1  -1   0   0   0   0   1   0   0   0   1   1   0;
     1   0  -1   1   0   0   0   0  -1   0   0   0   1   1   0;
     1   0   1  -1   0   0   0   0  -1   0   0   0   1   1   0;
     1   0   1   1   0   0   0   0   1   0   0   0   1   1   0;
     1  -1   0   0  -1   0   0   1   0   0   0   1   0   0   1;
     1  -1   0   0   1   0   0  -1   0   0   0   1   0   0   1;
     1   1   0   0  -1   0   0  -1   0   0   0   1   0   0   1;
     1   1   0   0   1   0   0   1   0   0   0   1   0   0   1;
```

```
1 -1  0 -1  0  0  1  0  0  0  0  1  0  1  0;
1 -1  0  1  0  0 -1  0  0  0  0  1  0  1  0;
1  1  0 -1  0  0 -1  0  0  0  0  1  0  1  0;
1  1  0  1  0  0  1  0  0  0  0  1  0  1  0;
1 -1 -1  0  0  1  0  0  0  0  0  1  1  0  0
1 -1  1  0  0 -1  0  0  0  0  0  1  1  0  0;
1  1 -1  0  0 -1  0  0  0  0  0  1  1  0  0;
1  1  1  0  0  1  0  0  0  0  0  1  1  0  0;]
```

% We write the design matrix of the Box and Behnken design for the 4 variable
% system

% 1st column is the constant "1" and 2nd, 3rd, 4th and 5th columns
% represent the variables in their coded forms while columns 6,7,8,9,10,11
% represent the product of variables with one another sequentially

% The columns12, 13, 14 and 15 show the squared values of each of the
% individual variables

```
Y = [0.205
     0.405
     0.19
     0.28
     0.26
     0.38
     0.315
     0.36
     0.17
     0.438
     0.24
     0.37
     0.257
     0.36
     0.2
     0.35
     0.23
     0.482
     0.417
     0.38
     0.15
     0.34
     0.42
     0.48]
```

% The Y matrix gives the list of responses obtained from experimentation at
% the specified conditions correspondingly

```
A = X\Y
```
% This command makes one fits the best possible set of
values for the matrix

```
% A which represents the coefficients that determine the
quadratic fit we
% intended
```

Results

Now, having run the code, the system is modelled as follows:

$$z = 0.3408 + 0.0357x_1 + 0.0239x_2 + 0.0394x_3 + 0.059x_4 - 0.0325x_1x_2$$

$$-0.0722x_1x_3 + 0.0117x_1x_4 - 0.0345x_2x_3 - 0.0187x_2x_4$$

$$-0.0275x_3x_4 + 0.0179x_1^2 - 0.0145x_3^2 - 0.0451x_4^2$$

where:
x_1 represents variable 1
x_2 represents variable 2
x_3 represents variable 3
x_4 represents variable 4

Thus, one successfully models the system using the Box and Behnken design of experimental approach.

Determination of maxima

This equation is differentiated and the maxima are evaluated to decide the maxima of the function. Four differential equations are written by differentiating the equation with respect to each of the variables. Then they are simultaneously solved to yield the point where this function achieves its maximum value. It could also be the minimum value but this can be checked by simply comparing that value with the value of the function at some other point.

This yields the point [25.26 −8.50 11.224 6.41], at a fitness value of 37.69 after 100 generations at the same settings of parameters as below.

This is not a practically feasible point, the optimum got is not something that can be used in reality as the values of the variables lie well outside that of the operable range of the variables, which is [−1 1].

Optimization within experimental limits

The goal is to optimize this system within the experimental limits. GA couples with the Box-Behnken design approach to come up with the optimal point of operation.

The above-mentioned model is the fitness function for the GA and to optimize it under the constraints of the values of the variables being within −1 and 1. The genetic algorithm optimization is done using the GA toolbox of MATLAB.

This optimization yields a fitness function value of 0.49736

The parameter values are as follows:

Number of generations = 13
Population size = 20 individuals
Scattered or random type crossover

Elite count used is 2
Crossover fraction used is 0.8
The simulation takes 1.1 s

Values of parameter come after multiple trials. This leads one to believe that values would most likely converge on the best value of the fitness function. The fitness function used is a negative of the z-function mentioned before. The stopping criterion used is either 100 generations or precision of the GA increment falling below that of a given value. The value of the variables at which this value was achieved was given by [1 1 −1 0.893]. Thus, the optimization is as per GA procedure.

Optimization with integer constraints

Therefore, as the next step, we perform optimization using GA with a constraint that the values of the parameters can only be integer values (−1, 0, 1). This becomes a very realistic case since we know that the system can definitely be operated at these conditions.

Running the simulations with the above mentioned integer constraint, data are as follows:

Fitness function value = 0.473

The parameter values are:

Number of generations = 7
Population size = 20 individuals
Scattered or random type crossover
Elite count used is 2
Crossover fraction used is 0.8
The simulation takes 0.6 s

The point at which this optima is reached is [1 1 −1 1].

Results

The point at which this optima is reached is [1 1 -1 1]. This point is not a part of the Box-Behnken design of experiments and is an extrapolation made using the trends modelled from the Box-Behnken experiment. The experimenter compares this predicted result with that obtained from the experiment.

This point compares with that obtained from experimentation using the Box-Behnken design of experiments. It is seen that even from experimentation the point that has the maximum optima is this point. The response got from our optimization is 0.497, while that got through experimentation is 0.485. If there are small errors due to the modelling, regression of data for the prediction is not accurate. The point obtained from the GA-coupled Box and Behnken optimization is going to perform much better than the one got from Box and Behnken alone.

Example 7.7: System modelled using the Box-Behnken design for five variables

The Box-Behnken design for five variables with responses is shown in Table 7.5.

TABLE 7.5

The Box-Behnken Design for Five Variables with Responses

Experiment Number	Variable 1	Variable 2	Variable 3	Variable 4	Variable 5	Response
1	0	0	0	−1	−1	0.2
2	0	0	0	−1	1	0.25
3	0	0	0	1	−1	0.18
4	0	0	0	1	1	0.27
5	0	0	−1	0	−1	0.35
6	0	0	−1	0	1	0.3
7	0	0	1	0	−1	0.32
8	0	0	1	0	1	0.4
9	0	0	−1	−1	0	0.5
10	0	0	−1	1	0	0.28
11	0	0	1	−1	0	0.52
12	0	0	1	1	0	0.61
13	0	−1	0	0	−1	0.33
14	0	−1	0	0	1	0.31
15	0	1	0	0	−1	0.54
16	0	1	0	0	1	0.69
17	0	−1	0	−1	0	0.27
18	0	−1	0	1	0	0.32
19	0	1	0	−1	0	0.4
20	0	1	0	1	0	0.65
21	0	−1	−1	0	0	0.35
22	0	−1	1	0	0	0.47
23	0	1	−1	0	0	0.38
24	0	1	1	0	0	0.7
25	−1	0	0	0	−1	0.21
26	−1	0	0	0	1	0.35
27	1	0	0	0	−1	0.4
28	1	0	0	0	1	0.62
29	−1	0	0	−1	0	0.3
30	−1	0	0	1	0	0.44
31	1	0	0	−1	0	0.52
32	1	0	0	1	0	0.7
33	−1	0	−1	0	0	0.29
34	−1	0	1	0	0	0.41
35	1	0	−1	0	0	0.45
36	1	0	1	0	0	0.65
37	−1	−1	0	0	0	0.18
38	−1	1	0	0	0	0.33
39	1	−1	0	0	0	0.42
40	1	1	0	0	0	0.66

The optimization procedure is as per earlier discussion. The MATLAB code 5 is as follows:

Table MATLAB code-5.

```
X = [1 0  0  0 -1 -1 0 0 0 0 0 0 0 0  0 1 0 0 0 1 1;
     1 0  0  0 -1  1 0 0 0 0 0 0 0 0 -1 0 0 0 0 1 1;
     1 0  0  0  1 -1 0 0 0 0 0 0 0 0 -1 0 0 0 0 1 1;
     1 0  0  0  1  1 0 0 0 0 0 0 0 0  1 0 0 0 0 1 1;
     1 0  0 -1  0 -1 0 0 0 0 0 0 0 1  0 0 0 0 1 0 1;
     1 0  0 -1  0  1 0 0 0 0 0 0 0 -1 0 0 0 0 1 0 1;
     1 0  0  1  0 -1 0 0 0 0 0 0 0 -1 0 0 0 0 1 0 1;
     1 0  0  1  0  1 0 0 0 0 0 0 0 1  0 0 0 0 1 0 1;
     1 0  0 -1 -1  0 0 0 0 0 0 0 1 0  0 0 0 0 1 1 0;
     1 0  0 -1  1  0 0 0 0 0 0 0 -1 0 0 0 0 0 1 1 0;
     1 0  0  1 -1  0 0 0 0 0 0 0 -1 0 0 0 0 0 1 1 0;
     1 0  0  1  1  0 0 0 0 0 0 0 1 0  0 0 0 0 1 1 0;
     1 0 -1  0  0 -1 0 0 0 0 0 1 0 0  0 0 1 0 0 0 1;
     1 0 -1  0  0  1 0 0 0 0 0 -1 0 0 0 0 1 0 0 0 1;
     1 0  1  0  0 -1 0 0 0 0 0 -1 0 0 0 0 1 0 0 0 1;
     1 0  1  0  0  1 0 0 0 0 0 1 0 0  0 0 1 0 0 0 1;
     1 0 -1  0 -1  0 0 0 0 0 1 0 0 0  0 0 1 0 1 0;
     1 0 -1  0  1  0 0 0 0 0 0 -1 0 0 0 0 1 0 1 0;
     1 0  1  0 -1  0 0 0 0 0 0 -1 0 0 0 0 1 0 1 0;
     1 0  1  0  1  0 0 0 0 0 0 1 0 0  0 0 1 0 1 0;
     1 0 -1 -1  0  0 0 0 0 0 1 0 0 0  0 0 1 1 0 0;
     1 0 -1  1  0  0 0 0 0 0 -1 0 0 0 0 0 1 1 0 0;
     1 0  1 -1  0  0 0 0 0 0 -1 0 0 0 0 0 1 1 0 0;
     1 0  1  1  0  0 0 0 0 0 1 0 0 0  0 0 1 1 0 0;
     1 -1 0  0  0 -1 0 0 0 1 0 0 0 0  0 1 0 0 0 0 1;
     1 -1 0  0  0  1 0 0 0 -1 0 0 0 0 0 1 0 0 0 0 1;
     1  1 0  0  0 -1 0 0 0 -1 0 0 0 0 0 1 0 0 0 0 1;
     1  1 0  0  0  1 0 0 0 1 0 0 0 0  0 1 0 0 0 0 1;
     1 -1 0  0 -1  0 0 0 1 0 0 0 0 0  0 1 0 0 1 0;
     1 -1 0  0  1  0 0 0 -1 0 0 0 0 0 0 1 0 0 1 0;
     1  1 0  0 -1  0 0 0 -1 0 0 0 0 0 0 1 0 0 1 0;
     1  1 0  0  1  0 0 0 1 0 0 0 0 0  0 1 0 0 1 0;
     1 -1 0 -1  0  0 0 1 0 0 0 0 0 0  0 1 0 1 0 0;
     1 -1 0  1  0  0 0 -1 0 0 0 0 0 0 0 1 0 1 0 0;
     1  1 0 -1  0  0 0 -1 0 0 0 0 0 0 0 1 0 1 0 0;
     1  1 0  1  0  0 0 1 0 0 0 0 0 0  0 1 0 1 0 0;
     1 -1 -1 0  0  0 1 0 0 0 0 0 0 0  0 1 1 0 0 0;
     1 -1  1 0  0  0 1 0 0 0 0 0 0 0  0 1 1 0 0 0;
     1  1 -1 0  0  0 1 0 0 0 0 0 0 0  0 1 1 0 0 0;
     1  1  1 0  0  0 1 0 0 0 0 0 0 0  0 1 1 0 0 0;]
```

% We write the design matrix of the Box-Behnken design for 5 variables
% system

% 1st column is the constant "1" and 2nd, 3rd, 4th, 5th and 6 th columns
% represent the variables in their coded forms while columns 7,8,9,10,11,12
% represent the product of variables with one another sequentially
% The columns13 ,14, 15, 16 and 17 show the squared values of each of the

```
% individual variables

Y = [0.2
     0.25
     0.18
     0.27
     0.35
     0.3
     0.32
     0.4
     0.5
     0.28
     0.52
     0.61
     0.33
     0.31
     0.54
     0.69
     0.27
     0.32
     0.4
     0.65
     0.35
     0.47
     0.38
     0.7
     0.21
     0.35
     0.4
     0.62
     0.3
     0.44
     0.52
     0.7
     0.29
     0.41
     0.45
     0.65
     0.18
     0.33
     0.42
     0.66]

% The Y matrix gives the list of responses got through
experimentation at
% the specified conditions correspondingly

A = X\Y

% This command makes us fit the best possible set of values
for the matrix
% A which represents the coefficients that determine the
quadratic fit we
% intended
```

Now having run the code, we get the coefficients and using this we can represent the system by a model.

$$z = 0.24 + 0.01197x_1 + 0.1063x_2 + 0.0737x_3 + 0.0306x_4 + 0.0412x_5 - 0.15x_1x_2$$

$$+0.02x_1x_3 + 0.01x_1x_4 + 0.02x_1x_5 + 0.05x_2x_3 + 0.05x_2x_4 + 0.0425x_2x_5$$

$$+0.0775x_3x_4 + 0.0325x_3x_5 + 0.01x_4x_5 + 0.1508x_1^2 + 0.1567x_2^2$$

$$+0.1050x_3^2 + 0.0575x_4^2$$

Now, the model equation is differentiated and the maxima are evaluated to decide the maxima of the function. Four differential equations are written by differentiating the equation with respect to each of the variables. Then they are simultaneously solved to yield the point at which this function achieves its maximum value. It could also be the minimum value but this can be checked by simply comparing that value with the value of the function at some other point.

This yields the point [23.14 23.28 9.62 3.16 5.26], at a fitness value of 252.87 after 100 generations at the same parameter values as below.

But this is not a practically feasible point, the optimum got is not something that can be used in reality as the values of the variables lie well outside that of the operable range of the variables, which is [−1 1].

The goal is to optimize this system within the experimental limits. For this, GA couples with the Box-Behnken design approach to come up with the optimal point of operation.

We use the above-mentioned model as the fitness function for the GA and optimize it under the constraints of the values of the variables being within −1 and 1. The genetic algorithm optimization is done using the GA toolbox of MATLAB.

The GA is naturally programmed to minimize a function and not maximize it, so while writing the fitness function we edit it such that it is written as the negative of the actual fitness function.

This optimization yields a fitness function value of 13.4736

The parameter values are as follows:

Number of generations = 33
Population size = 20 individuals
Scattered or random type crossover
Elite count used is 2
Crossover fraction used is 0.8
The simulation takes 2 s

These parameter values selected after multiple trials leads to believe that these values would most likely converge on the best value of the fitness function. The fitness function used was a negative of the z-function mentioned before. The stopping criterion used was either 100 generations or precision of the GA increment falling below that of a given value.

The value of the variables at which this value was achieved was given by [1 0.997 0.998 0.997 0.998].

Running the simulations with the above mentioned integer constraint gives: Fitness function value = 1.243

The parameter values are as follows:

Number of generations = 37
Population size = 20 individuals
Scattered or random type crossover
Elite count used is 2
Crossover fraction used is 0.8
The simulation takes 1.9 s

The point at which this optima is reached is [1 1 1 1 1]. This point is not a part of the Box-Behnken design of experiments and is an extrapolation made using the trends modelled from the Box-Behnken experiment. So one compares this predicted result with that got from experiment.

This point compares with that got from experimentation using the Box-Behnken design of experiments. It is seen that even from experimentation the point that has the maximum optima is this point. The response got from our optimization is 1.243, while that got through experimentation is 0.69. Clearly, even if there are small errors due to the modelling – as one performs regression of data the prediction is not accurate, the point got from the GA-coupled Box and Behnken optimization is going to perform much better than the one got from simple Box and Behnken.

Exercises

7.1 What are the advantages of using genetic algorithm for optimizing a biological system instead of a response surface methodology design for the same system?

7.2 For a biological system with three variables, each being able to take two levels each, how many experiments have to be performed if factorial design is used? How would this compare (qualitatively) to using a completely independent systematic GA? (A set of experiments are performed and then the responses recorded and using this, the next set of experiments are chosen.)

7.3 Consider a biological system, which has to be modelled, but there are 27 variables that can affect the response of the system and seven of these can take three different levels while the remaining 20 of them can take two different levels. How many experiments have to be performed using a factorial design? How would you compare this (qualitatively) to using a completely systematic genetic algorithm-based optimization? Is your conclusion different from the previous question? If it is yes, then why is this?

7.4 Discuss the influence of the following factors on the convergence and efficiency of optimization achieved through genetic algorithm: mutation frequency, population size, crossover fraction, type of crossover, and use of binary or ternary bits for coding the population.

7.5 Consider a biological system, which is controlled by multiple experimental conditions. Let A, B, C, D, and E be the different variables that control the system and the response of the system can be mathematically modelled as follows:

$$Y = 0.9A^2 + 1.05B^2 + 2.1C^2 + 1.4D^2 + 9.7E^2 + 100.2A + 93B$$

$$+5.7C + 0.4D + 94E + 75AB + 67AC + 84AD + 2.1AE$$

$$+7.4BC + 6.1BD + 8.05BE + 3CD + 24CE + 56DE$$

$$+108ADE + 98ACD + 0.5BDE + 1.15ABC$$

For such a complicated system, it is almost impossible to calculate the solution by hand (even harder to find maxima/minima within a particular domain/limit). Using GA, calculate the absolute maximum productivity and minimum productivity of the system within the limits of −3 to 3 for each of the individual variables.

7.6 Consider a biological system (Table 7.6.1), which has been modelled using the factorial design, and this system has to be optimized for the best possible value within this domain. In a coded format, the factorial design has been performed and results are recorded as given below. Find the maximum response that can be achieved by this system within the domain of the factorial design experiment points using the GA toolbox in MATLAB or any other software that has the GA module.

TABLE 7.6.1

Design Matrix with the Response for Tartaric Acid Fermentation by *Gluconobacter* sp.

Experiment Number	X_1	X_2	X_3	X_4	X_5	Response
1	0	0	0	0	0	62.7
2	0	0	0	0	1	63.4
3	0	0	0	1	0	65.9
4	0	0	0	1	1	65.8
5	0	0	1	0	0	61.2
6	0	0	1	0	1	62.5

(Continued)

TABLE 7.6.1 (*Continued*)

Design Matrix with the Response for Tartaric Acid Fermentation by
Gluconobacter sp.

Experiment Number	X_1	X_2	X_3	X_4	X_5	Response
7	0	0	1	1	0	63.8
8	0	0	1	1	1	63.4
9	0	1	0	0	0	63.1
10	0	1	0	0	1	66.6
11	0	1	0	1	0	68.4
12	0	1	0	1	1	68.1
13	0	1	1	0	0	69.0
14	0	1	1	0	1	69.9
15	0	1	1	1	0	69.7
16	0	1	1	1	1	69.5
17	1	0	0	0	0	69.7
18	1	0	0	0	1	61.2
19	1	0	0	1	0	63.8
20	1	0	0	1	1	65.6
21	1	0	1	0	0	64.4
22	1	0	1	0	1	62.0
23	1	0	1	1	0	61.7
24	1	0	1	1	1	69.9
25	1	1	0	0	0	67.4
26	1	1	0	0	1	63.1
27	1	1	0	1	0	65.3
28	1	1	0	1	1	66.2
29	1	1	1	0	0	66.1
30	1	1	1	0	1	65.7
31	1	1	1	1	0	64.8
32	1	1	1	1	1	62.3

7.7 As a process engineer at a biochemical manufacturing plant,
you are asked to determine the maximum product that can be
produced within the limits of the equipment. For that purpose,
you start to perform experiments on the system, which can be
modelled by four different parameters (variables) each taking
(2 + 1) different levels. The results of the experimentation are in
Table 7.7.1. A central cubic design (rotatable design) was used.
What would be the operating point of the system for maximum
product synthesis? (The system boundaries are constrained by the
equipment limits—which are the values at which the experiments
are performed.)

TABLE 7.7.1

Experimental Matrix with Responses

Experiment Number	X_1	X_2	X_3	X_4	Response
1	0	0	0	0	11.4
2	0	0	0	1	11.1
3	0	0	1	0	12.3
4	0	0	1	1	13.7
5	0	1	0	0	12.9
6	0	1	0	1	11.8
7	0	1	1	0	13.2
8	0	1	1	1	11.4
9	1	0	0	0	11.6
10	1	0	0	1	13.1
11	1	0	1	0	12.7
12	1	0	1	1	12.5
13	1	1	0	0	11.9
14	1	1	0	1	12.4
15	1	1	1	0	11.1
16	1	1	1	1	10.9
17	2	0	0	0	13.2
18	−2	0	0	0	11.7
19	0	2	0	0	12.9
20	0	−2	0	0	12.7
21	0	0	2	0	12.7
22	0	0	−2	0	11.1
23	0	0	0	2	11.6
24	0	0	0	−2	12.4
25	0	0	0	0	13.6
26	0	0	0	0	13.6
27	0	0	0	0	13.5
28	0	0	0	0	13.5

7.8 You are a research scholar working on determining if a particular bioprocess is economically feasible or not. For this, you have to first determine the maximum amount of product that can be produced for a given feed, which is determined mainly by the experimental conditions that is maintained (Table 7.8.1). For this purpose form theory, you have arrived at two very important variables that affect your productivity significantly. Your assistant who is helping with your project has taken it up for a couple of weeks, as you were ill. During this time, he has performed experiments using a CCD orthogonal design matrix and recorded the observations as below. Unfortunately, he does not have the expertise to interpret his results.

TABLE 7.8.1

Design Matrix with the Response for Problem 7.8

Experiment Number	Variable A	Variable B	Response Recorded
1	0	0	468.3
2	0	1	457.1
3	1	0	473.9
4	1	1	458.1
5	1.414	0	455
6	−1.414	0	443.8
7	0	1.414	461.7
8	0	−1.414	458.9
9	0	0	471
10	0	0	470.6
11	0	0	471.2
12	0	0	471.1

Having recovered from the flu that you got you have bigger problems to worry about. You have to submit your final report in a few days – looks like genetic algorithm is the easiest and most efficient way out for you.

7.9 The response of a bioprocess is controlled by three variables each of which can be maintained at levels, −1 or 0 or 1, each. However, performing one single experiment is very costly and hence, an elaborate factorial design of experiments cannot be performed for this system. Therefore, the experimenter has decided that the Box and Behnken design (assuming that 0 is the middle value, which is neutral) is the way to go. Using these experiments, the results were recorded in Table 7.9.1.

Determine the maximum production that can be achieved from this system within the experimental domain with the help of genetic algorithm (after fitting an appropriate second degree polynomial on this grid).

7.10 Consider the system defined in Problem 7. 6. Now, we shall evaluate the practical advantages of directly applying GA versus that for applying a response surface methodology (in this case factorial design) and then GA on top of it. The cost of performing one such experiment is close to $1000. (This is not an unusually high amount for a bioprocess, considering the very high capital costs). Determine the cost for performing a factorial design and compare that with the cost that we need if we use a genetic algorithm for the same. In evaluating the GA cost, neglect the computational cost as it is very

TABLE 7.9.1

Experimental Results for Problem 7.9

Experiment Number	Variable 1	Variable 2	Variable 3	Response
1	−1	−1	0	4.56
2	−1	1	0	4.64
3	1	−1	0	4.58
4	1	1	0	4.43
5	0	−1	−1	4.39
6	0	−1	1	4.61
7	0	1	−1	4.70
8	0	1	1	4.53
9	−1	0	−1	4.54
10	−1	0	1	4.73
11	1	0	−1	4.59
12	1	0	1	4.60

negligible compared to experimentation cost. (Hint: choose a random starting population, take out the responses from the factorial table, and compute the next generations and so on until you reach target productivity, then compute total cost for distinct experiments performed this way). What does this tell you about GA?

7.11 For the case of both rotatable and orthogonal CCD, which is the most commonly used due to its obvious advantages of being both orthogonal and rotatable. Table 7.11.1 gives the grid to carry out experiments.

If we use this as our starter for the genetic algorithm and then record all the responses, then apply the genetic algorithm as mentioned in the flowchart. Get the solution. Apply the principle of a GA-coupled hybrid design.

TABLE 7.11.1

Grid for Rotatable and Orthogonal CCD

Experiment Number	Variable 1	Variable 2	Variable 3	Variable 4
1	−1	−1	−1	−1
2	−1	−1	−1	0
3	−1	−1	−1	1
4	−1	−1	0	−1
5	−1	−1	0	0
6	−1	−1	0	1
7	−1	−1	1	−1

(Continued)

TABLE 7.11.1 (*Continued*)

Grid for Rotatable and Orthogonal CCD

Experiment Number	Variable 1	Variable 2	Variable 3	Variable 4
8	−1	−1	1	0
9	−1	−1	1	1
10	−1	0	−1	−1
11	−1	0	−1	0
12	−1	0	−1	1
13	−1	0	0	−1
14	−1	0	0	0
15	−1	0	0	1
16	−1	0	1	−1
17	−1	0	1	0
18	−1	0	1	1
19	−1	1	−1	−1
20	−1	1	−1	0
21	−1	1	−1	1
22	−1	1	0	−1
23	−1	1	0	0
24	−1	1	0	1
25	−1	1	1	−1
26	−1	1	1	0
27	−1	1	1	1
28	0	−1	−1	−1
29	0	−1	−1	0
30	0	−1	−1	1
31	0	−1	0	−1
32	0	−1	0	0
33	0	−1	0	1
34	0	−1	1	−1
35	0	−1	1	0
36	0	−1	1	1
37	0	0	−1	−1
38	0	0	−1	0
39	0	0	−1	1
40	0	0	0	−1
41	0	0	0	0
42	0	0	0	1
43	0	0	1	−1
44	0	0	1	0
45	0	0	1	1
46	0	1	−1	−1
47	0	1	−1	0

(*Continued*)

TABLE 7.11.1 (*Continued*)

Grid for Rotatable and Orthogonal CCD

Experiment Number	Variable 1	Variable 2	Variable 3	Variable 4
48	0	1	−1	1
49	0	1	0	−1
50	0	1	0	0
51	0	1	0	1
52	0	1	1	−1
53	0	1	1	0
54	0	1	1	1
55	1	−1	−1	−1
56	1	−1	−1	0
57	1	−1	−1	1
58	1	−1	0	−1
59	1	−1	0	0
60	1	−1	0	1
61	1	−1	1	−1
62	1	−1	1	0
63	1	−1	1	1
64	1	0	−1	−1
65	1	0	−1	0
66	1	0	−1	1
67	1	0	0	−1
68	1	0	0	0
69	1	0	0	1
70	1	0	1	−1
71	1	0	1	0
72	1	0	1	1
73	1	1	−1	−1
74	1	1	−1	0
75	1	1	−1	1
76	1	1	0	−1
77	1	1	0	0
78	1	1	0	1
79	1	1	1	−1
80	1	1	1	0
81	1	1	1	1
82	3	0	0	0
83	−3	0	0	0
84	0	3	0	0
85	0	−3	0	0
86	0	0	3	0
87	0	0	−3	0
88	0	0	0	3

(Continued)

TABLE 7.11.1 (*Continued*)

Grid for Rotatable and Orthogonal CCD

Experiment Number	Variable 1	Variable 2	Variable 3	Variable 4
89	0	0	0	−3
90	0	0	0	0
91	0	0	0	0
92	0	0	0	0
93	0	0	0	0
94	0	0	0	0
95	0	0	0	0
96	0	0	0	0
97	0	0	0	0
98	0	0	0	0
99	0	0	0	0
100	0	0	0	0
101	0	0	0	0
102	0	0	0	0
103	0	0	0	0
104	0	0	0	0
105	0	0	0	0
106	0	0	0	0
107	0	0	0	0
108	0	0	0	0
109	0	0	0	0
110	0	0	0	0
111	0	0	0	0
112	0	0	0	0
113	0	0	0	0
114	0	0	0	0
115	0	0	0	0
116	0	0	0	0
117	0	0	0	0
118	0	0	0	0
119	0	0	0	0
120	0	0	0	0
121	0	0	0	0

References

1. Melanie M, (Ed.), *An Introduction to Genetic Algorithms*, MIT Press, 1998.
2. Eiben A E et al., Genetic algorithms with multi-parent recombination. PPSN III: Proceedings of the International Conference on Evolutionary Computation. The Third Conference on Parallel Problem Solving from Nature: 78–87, 1994.

3. Taherdangkoo M, Paziresh, M, Yazdi M and Hadi BM, An efficient algorithm for function optimization: modified stem cells algorithm, *Central European Journal of Engineering* **3** (1): 36–50, 2012.
4. Wolpert DH and Macready WG, No free lunch theorems for optimisation. Santa Fe Institute, SFI-TR-05-010, Santa Fe, 1995.
5. Ting C-K, On the mean convergence time of multi-parent genetic algorithms without selection, *Advances in Artificial Life*: 403–412, 2005.
6. Akbari Z, A multilevel evolutionary algorithm for optimizing numerical functions, *IJIEC* **2**: 419–430, 2011.
7. Eshelman LJ, The CHC adaptive search algorithm: How to have safe search when engaging in nontraditional genetic recombination, in GJ E Rawlins (Ed.), *Proceedings of the First Workshop on Foundations of Genetic Algorithms*, pp. 265–283. Morgan Kaufmann, 1991.
8. Gaitonde VN, Karnik SR, Siddeswarappa B and Achyutha BT, Integrating Box-Behnken design with genetic algorithm to determine the optimal parametric combination for minimizing burr size in drilling of AISI 316L stainless steel, Springer-Verlag, London Limited, 2007.
9. Panda T, (Ed.), *Bioreactors Analysis and Design*, Tata McGraw-Hill Education Pvt. Ltd., New Delhi, India, 2011.

Further Reading

1. Goldberg DE, *Genetic Algorithms in Search, Optimization, and Machine Learning*, Pearson Education Inc., and Dorling Kinderseley Publishing Inc., New Delhi, India, 2013.

Appendix

A.1 Standard Tables for Data to Calculate Statistic Parameters

Percentage points of t-distributions and F-distributions with different degrees of freedom are available in any standard textbooks of statistics.[1]

A.2 Complex Algorithm of Box

The complex algorithm of Box,[2] later modified by Kuester and Mize,[3] is a sequential search technique that has proven to be effective while solving problems with non-linear inequality constraints. This procedure finds the maximum of a multi-variable, non-linear function subject to non-linear inequality constraints.

In the present work, the complex algorithm of Box was employed to maximize the following function:

$$\hat{Y}_i = \beta_0 + \beta_1 X_1 + \beta_2 X_2 + \beta_3 X_3 + \beta_{11} X_1^2 + \beta_{22} X_2^2 + \beta_{33} X_3^2 + \beta_{12} X_1 X_2 + \beta_{13} X_1 X_3 + \beta_{23} X_2 X_3$$

subject to the following constraints:

$$-\alpha \le X_i \le +\alpha \; ; i = 1, 2, 3, \dots$$

where:
   α is the distance of the axial point from the design centre in the central composite experimental design

The program listing for the complex algorithm of Box is given in Section A.3. This is modified based on the concept of Box[2] which is further modified by Kuester and Mize.[3] Further improvement and modification of the program is possible. This is one of the examples. Further information is described by Panda.[4]

A.3 Modified Algorithm of Box, Kuester, and Mize

The program is written to maximize the response.

```
CC      MAIN LINE PROGRAM FOR COMPLEX ALGORITHM OF BOX
        DIMENSION X (6,4), R (6,3), F (6), G (4), H (4), XC (3)
        INTEGER GAMMA
C
        NI = 50
        NO = 60
        OPEN (NI, FILE = ' INP1. DAT ')
        OPEN (NO, FILE = ' OUT1. DAT ')
C       READ (NI, 001) N, M, K, ITMAX, IC, IPRINT
001     FORMAT (815)
        READ (NI, *) ALPHA, BETA, GAMMA
c002    FORMAT (2E10.4, 15)
        DELTA = 0.0001
        READ (NI, *) (X(1, J), J = 1, N)
c       READ (NI, 004) (X(1, J), J = 1, N)
c003    FORMAT (16F5.4)
100     CONTINUE
C
        WRITE (NO, 010)
010     FORMAT (1H1, //, 18X, 24HCOMPLEX PROCEDURE OF BOX)
        WRITE (NO, 018)
018     FORMAT (//, 2X, 10HPARAMETERS)
        WRITE (NO, 011) N, M, K, TTMAX, IC, ALPHA, BETA, GAMMA,
        DELTA
011     FORMAT (//, 2X, 4HN = , 12, 3X, 4HK =, 12, 2K, BHITMAX =,
     1  14, 2X, 5HIC =, 12, //, 2X, BHALPHA =, F5, 3, 5X, 7HBETA =,
        F10.5, 3X,
     2  8HGAMMA =, 12, 3X, BHDELTA =, F6, 5)
40      WRITE (NO, 012)
012     FORMAT (//, 2X, 14HRANDOM NUMBER)
        DO 200 J = 2, K
        WRITE (NO, 013) (J, I, R(J, I), I = 1, N)
013     FORMAT (/, 3(2X, 2HR( ,12, 1H, 12, 4H) = , F6, 4, 2X))
200     CONTINUE
```

```
C
50      CALL CONSX (N, M, K, ITMAX, ALPHA, BETA, GAMMA,
        DELTA, X, R,
    1   F, IT, IEV2, NO, G, H, XC, IPRINT)
C
        IF (IT–ITMAX) 20, 20, 30
20      WRITE (NO, 014) F (IEV2)
014     FORMAT (///, 2X, 30HFINAL VALUE OF THE FUNCTION =
        E20.8)
        WRITE (NO, 015)
015     FORMAT (///, 2X, 14HFINAL X VALUES)
        D 300 J = 1, N
        WRITE (NO, 016) J, X(IEV2, J)
016     FORMAT (///, 2X, 2HX( , 12, 4H) = , F20.8)
300     CONTINUE
        GO TO 999
C
30      WRITE (NO, 017) ITMAX
017     FORMAT (///, 2X, 38HTHE NUMBER OF ITERATIONS HAS
        EXCEEDED ,
    1   14, 10X, 18HPROGRAM TERMINATED)
999     STOP
        END
        SUBROUTINE CONSX (N,M,K, ITMAX, ALPHA, BETA,
        GAMMA, DELTA, X, R, F, IIT, IEV2, NO, G, H, XC, IPRINT)
C       COORDINATES SPECIAL PURPOSE SUBROUTINES
C
C
C
C       DIMENSION X(K, M), R(K, N), F(K), G(M), H(M), XC(N)
C
C
        IT = 1
        KODE = 0
        IF (M–N) 20, 20, 10
10      KODE = 1
        CONTINUE
```

```
      DO 40 II = 2K
      DO 30 J = 1, N
30    X(II, J) = 0.0
40    CONTINUE
C
C
C     CALCULATE COMPLEX POINTS AND CHECK AGAINST
      CONSTRAINTS

      DO 65 II = 2, K
      DO 50 J = 1, N
      I = II
      CALL CONST (N, M, K, X, G, H, I)
      X(II, J) = G(J) + R(II, J)*(H(j) – G(J))
50    CONTINUE
      K1 = II
      CALL CHECK (N, M, K, X, G, H, I, KODE, XC, DELTA, K1)
      IF (II–2) 51, 51, 55
51    IF (IPRINT) 52, 65, 52
52    WRITE (NO, 018)
018   FORMAT (//, 2X, 2HX (, 12, 1H, ,12, 4H) = , 1PE13.6
      I0 = 1
      WRITE (NO, 019) (IO.J.X.(IO,J) ,J = 1 ,N)
019   FORMAT(/, 3 (2X, 2HX (,12, 1H, , 12, 4H) = , 1PE13.6
55    IF (IPRINT) 56, 65, 56
56    WRITE (NO, 019) (II, J, X(II, J), J = 1, N)
65    CONTINUE
      K1 = K
      DO 70 I = 1, K
      CALL FUNC (N, M, K, X, F, I)
70    CONTINUE
      KCOUNT = 1
      IA = 0
CC
CC    FIND POINT WITH LOWEST FUNCTION VALUE
```

```
        IF (IPRINT) 72, 80, 72
72      WRITE (NO, 021)
021     FORMAT (/, 2X, 22HVALUES OF THE FUNCTION)
        WRITE (NO, 022) (J, F(J), J = 1, K)
022     FORMAT (/, 3(2X, 2HF( ,I2, 4H ( , I2, 4H) = , 1PE13.6))
80      IEV = 1
        DO 100 ICM = 2, K
        IF (F(IEV1) – F(ICM)) 100, 100, 90
90      IEV1 = ICM
100     CONTINUE
C
C
C       FIND POINT WITH HIGHEST FUNCTION VALUE
        IEV1 = 1
        DO 120 ICM = 2, K
        IF (F(IEV2) – F(ICM)) 110, 110, 120
110     IEV2 = ICM
120     CONTINUE
C
C
C
C       CHECK CONVERGENCE CRITERIA
        IF ( F(IEV2) – (F(IEV1) + BETA)) 140, 130, 130
130     KOUNT = 1
        GO TO 150
140     KOUNT = KOUNT + 1
        IF (KOUNT–GAMMA) 150, 240, 240
C
C       REPLACE POINT WITH LOWEST FUNCTION VALUE
C
150     CALL CENTR (N, M, K, IEV1, I, XC, X, K1)
        DO 160 JJ = 1, N
160     X(IEV1, JJ) = (1.0 + ALPHA)*(XC(JJ)) – ALPHA*(X(IEV1, JJ))
        IE = IEV1
```

```
        CALL CHECK (N, M, K, X, G, H, I, KODE, XC, DELTA, K1)
        CALL FUNC (N, M., K, X, F, I)
C
C
C       REPLACE WITH NEW POINT IF IT REPEATES AS LOWEST
        FUNCTION VALUE
170     IEV2 = 1
        DO 190 ICM = 2, K
        IF (F (IEV2) – F(ICM)) 190, 190, 180
180     IEV2 = ICM
190     CONTINUE
        IF (F (IEV2) – F(IEV1) 220, 200, 220
200     DO 210 JJ = 1, N
        X(IEV1, JJ) = (X(IEV1, JJ))/2.0
210     CONTINUE
        I = IEV1
        CALL CHECK (N, M, K, X, G, H, I, KODEMXC, DELTA, K1)
        CALL FUNC (N, M, K, X, F, I)
        GO TO 170
220     CONTINUE
        IF (IPRINT) 230, 228, 230
230     WRITE (NO, 023) IT
023     FORMAT (//, 2X, 17HITERATION NUMBER, 115)

        WRITE (NO, 024)
024     FORMAT (/, 2X, 30HCOORDINATES OF THE CORRECTED
        POINTS)
        WRITE (NO, 019) (IEV1, JC, X(IEV1, JC), JC = 1, N
        WRITE (NO, 021)
        WRITE (NO, 022) (I, F(I), I = 1, K)
        WRITE (NO, 025)
025     FORMAT (/, 2X, 2HX (, 12, 6H, C) = , 1PE14.6, 4X))
        WRITE (NO, 026) (JC, XC(JC), JC = 1, N)
026     FORMAT (/, 3(2X, 2HX (, 12, 6H, C) =, 1PE14.6, 4X))
228     IT = IT + 1
```

```
        IF (IT – ITMAX) 80, 80, 240
240     RETURN
        END

        SUBROUTINE CHECK (N, M, K, X, G, H, I, KODE, XC, DELTA, K1)
C
CC
C       ARGUMENTS LISTS

10      DIMENSION X(K, M), G(M), H(M), XC(N)
        KT = 0
        CALL CONST (N, M, K, X, G, H, I)
C
C       CHECK AGAINST EXPLICIT CONSTRAINTS

        DO 50 J = 1, N
        IF ( X(I, J) – G(J)) 20, 20, 30
20      X(I, J) = G(J) + DELTA
        GO TO 50

30      IF (H(J) – X(I, J)) 40, 40, 50
40      X(I, J) = H(J) – DELTA
50      CONTINUE
        IF (KODE) 110, 110, 60
C
C       CHECK AGAINST THE IMPLICIT CONSTRAINTS
C

60      NN = N + 1
        DO 100 J = NN, M
        CALL CONST (N, M, K, X, G, H, I)
        IF (X(I, J) – G(J)) 80, 70, 70
70      IF (H(J) – X(I, J)) 80, 100, 100
80      IEV1 = 1
        KT = 1
        CALL CENTR (N, M, K, IEV1, I, XC, X, K1)
```

```
            DO 90 JJ = 1, N
            X(I, JJ) = (X(I, JJO + XC(JJ) / 2.0
90          CONTINUE
100         CONTINUE
            IF (KT) 110, 110, 10
110         RETURN
            END

            SUBROUTINE CENTR (N, M, IEV1, I, XC, X, K1)
            DIMENSION X(K, M), XC(N)
C
            DO 20 J = 1, N
            XC(J) = 0.0
            DO 10 IL = 1, K1
10          XC(J) = XC(J) + X(IL, J)
            RK = K1
20          XC(J) = (XC(J) – X(IEV1, J)) / RK – 1.0
            RETURN
            END

            SUBROUTINE FUNC (N, M, K, X, F, I)
            DIMENSION X(K, M), F(K)
C
C           POST OFFICE TEST PROBLEM
            F(I) = 0.3663 – 0.0139* X(I, 1) + 0.0622*X(I, 2) + 0.0126*X(I, 3) –
    1       0.0064*(X(I, 1)**2) + 0.0194* (X(I, 2)**2) – 0.0150* (X(I, 3)**2)
    2       + 0.0319*X(I, 1)*X(I, 2) + 0.0096*X(I, 2)*X(I, 3)
            RETURN
            END

            SUBROUTINE FUNC (N, M, K, X, G, H, I)
            DIMENSION X(K, M), G(M), H(M)
C
C
```

C   POST OFFICE TEST PROBLEM

$G(1) = -2.0$
$H(1) = 2.$
$G(2) = -2.0$
$H(2) = 2.$
$G(3) = -2.0$
$H(3) = 2.$
$G(4) = -2.0$
$H(4) = 2.$
C $X(I, 4) = X(I, 1) = 2.0*X(I, 2) + 2.0*X(I, 3)$
C

RETURN
END

A.4 MATLAB Program to Minimize the Distance Function

The program is written to minimize the distance function $\rho[Y,\phi]$.

```
% program to minimize the distance function
clear

x0 = [0 0 0]
x = min ('min', x0)
function f = min (x)

% x₁ = slant age
% x₂ = inoculum age
% x₃ = number of cells
f1 = [6.05 + 0.03 x₁ + 0.012 x₂ + 0.1 x₃ + 0.21 x₁² + 0.27 x₂² +
0.25 x₃² + 0.062 x₁x₂ - 0.038 x₁x₃ + 0.14 x₂x₃ - 6.189] ^2/
(6.189) ^2
f2 = 1.879E-03 - 7.604E-05 x1-3.169E-06 x2 - 1.312E-05 x3
- 1.499E-04 x₁² - 1.083E-04 x₂² - 1.872E-04 x₃² + 1.147E-04 x₁x₂
- 3.806E-06 x₁x₃ + 7.112E-05 x₂x₃ - 0.001878]^2/(0.001878)^2
f = (f1 + f2) ^ 1/2
end
```

A.5 Modified Algorithm for Rosenbrock Program

The program is written for maximization of a variable.[5]

```
      C
      C     MAIN LINE PROGRAM FOR ROSENBROCK
      C
            DIMENSION X(2), E(2), V(2,2), SA(2), D(2), G(2), H(2), AL(2),
            PH(2), A(2,2), B(2,2), BX(2), DA(1), VV(2,2), EINT(2), VM(2)
            COMMON COUNT
            INTEGER P
            INTEGER PR
            INTEGER R
            INTEGER C
            REAL LC
      C
            NI = 50
            NO = 66
      C
            READ (NI, 001), M, P, L, LOOPY, PR, ND, NDATA, NSTEP
  001       FORMAT (8I5)
  10        DO 100 K = 1, P
            READ (NI, 002) X (K)
  002       FORMAT (1E10.4)
  100       CONTINUE
            DO 200 J = 1, P
            READ (NI, 002) E (J)
  200       CONTINUE
            WRITE (NO, 013)
  013       FORMAT (1H1, 10X, 30HROSENBROCK PROCEDURE)
      C
            IF (ND–1) 30, 20, 30
  20        DO 300 KA = 1, NDATA
            READ (NI, 002) DA (KA)
  300       CONTINUE
  30        LAP = PR–1
            LOOP = 0
```

```
        ISW = 0
        INIT = 0
        KOUNT = 0
        TERM = 0.0
        DELY = 1.0E–10
        F1 = 0.0
        NPAR = NDATA
        N = L
        DO 40 K = 1,L
40      AL(K) = (CH(X, DA, N, NPAR, K)–CG(X, DA, N, NPAR,
        K))*.0001
        DO 60 I = 1,P
        DO 60 J = 1,P
        V(I,J) = 1.0
        IF(I–J) 60, 61,60
61      V(I,J) = 1.0
60      CONTINUE
        DO 65 KK = 1,P
        EINT (KK) = E(KK)
65      CONTINUE
C
C
100     DO 70 J = 1,P
        IF(NSTEP.EQ.0)E(J) = ENT(J)
        SA(JJ) = 2.0
70      D(J) = 0.0
        FBEST = F1
80      I = 1
        IF (INIT.EQ.0) GO TO 120
90      DO 110 K = 1,P
110     X(K) = X(K) + E(I)*V(I,K)
        DO 50 K = 1,L
50      H(K) = F0
C
C
```

```
120    F1 = F(X, DA, N, NPAR)
       F1 = M*F1
       IF (ISW.EQ,0) F0 = F1
       ISW = 1
       IF (ABS(FBEST–F1)–DELY) 122, 122, 125
122    TERM = 1.0
       GO TO 450
125    CONTINUE
C
C
       J = 1
C
130    XC = CX(X, DA, N, NPAR, J)
       LC = CG(X, DA, N, NPAR, J)
       UC = CH(X, DA, N, NPAR, J)
       IF (XC.LE.LC) GO TO 420
       IF (XC.GE.UC) GO TO 420
       IF (F1.LT.F0) GO TO 420
       IF (XC.LT.LC + AL(J)) GO TO 140
       IF (XC.GT.UG–AL(J)) GO TO 140
       H(J) = F0
       GO TO 210
C
C
140    CONTINUE
C
       BW = AL(J)
C
       IF (XC.LE.LC.OR.UC.LE.XC) GO TO 150
       IF (LC.LT.XC.AND.XC.LT.LC + BW) GO TO 160
       IF (UC–BW.LT.XC.AND.XC.LT.UC) GO TO 170
       PH(J) = 1.0
       GO TO 210
```

```
C
C
150    PH(J) = 0.0
       GO TO 190
160    PW = (LC + BW–XC)/BW
       GO TO 180
170    PW = (XC–UC + BW)/BW
180    PH(J) = 1.0–3.0*PW + 4.0*PW*PW–2.0 + PW*PW*PW
C
190    F1 = H(J) + (F1–H(J)*PH(J)
C
210    CONTINUE
       IF (J.EQ.L) GO TO 220
       J = J + 1
       GO TO 130
C
220    INIT = 1
       IF (F1.LT.F0) GO TO 420
       D(I) = D(I) + E(I)
       E(I) = 3.0*E(I)
       F0 = F1
       IF (SA(I).GE.0.5) SA(I) = 1.0
C
230    DO 240 JJ = 1,P
       IF (SA(JJ).GE.0.5) GO TO 440
240    CONTINUE
C
C      AXES ROTATION
C
       DO 250 R = 1,P
       DO 250 C = 1,P
250    VV(C,R) = 0.0
       DO 260 R = 1,P
```

```
        KR = R
        DO 260 C–1,P
        DO 265 K = KR,P
265     VV(R,C) = D(K)*V(K,C) + VV(R,C)
260     B(R,C) = VV(R,C)
        BMAG = BMAG + B(1,C)*B(1,C)
280     CONTINUE
        BMAG = SQRT (BMAG)
        BX(1) = BMAG
        DO 310 C = 1,P
310     V(1, C) = B(1,C)/BMAG
C
        DO 390 R = 2,P
C
        IR = R–1
        DO 390 C = 1,P
        SUMVM = 0.0
        DO 320 KK = 1, IR
        SUMAV = 0.0
        DO 330 KJ = 1,P
330     SUMAV = SUMAV + VV(R,KJ)*V(KK,KJ)
320     SUMAV = SUMAV*V(KK,C) + SUMVM
390     B(R,C) = VV(R,C)–SUMVM
        DO 340 R = 2,P
        BBMAG = 0.0
        DO 350 K = 1,P
350     BBMAG = BBMAG + B(R,K)*B(R,K)
        BBMAG = SQRT(BBMAG)
        DO 340 C = 1,P
340     V(R,C) = B(R,C)/BBMAG
        LOOP = LOOP + 1
        LAP = LAP + 1
        IF (LAP.EQ.PR) GO TO 450
        GO TO 1000
```

```
C
420   IF (INIT.EQ.0) GO TO 450
      DO 430 IX = 1,P
430   X(IX) = X(IX)–E(I)*V(I,IX)
      E(I) = –0.5*E(I)]
      IF (SA(I).LT.1.5) SA(I) = 0.0
      GO TO 230
C
440   CONTINUE
      IF (I.EQ.P)GO TO 80
      I = I + 1
      GO TO 90
C
450   WRITE (NO, 003)
003   FORMAT (///, 2X, 5HSTAGE, 8X, 8HFUNCTION, 12X,
      8HPROGRESS, 9X, 16HLATERAL PROGRESS)
      WRITE (NO, 004) LOOP, F0, BMAG, BBMAG
004   FORMAT (1H, I5, 3E20.8)
      WRITE (NO, 014) KOUNT
014   FORMAT (/, 2X,33HNUMBER OF FUNCTION
      EVALUATIONS = , I8)
      WRITE (NO, 005)
005   FORMAT (/, 2X, 25HVALUES OF XAT THIS STAGE)
C     PRINT CURRENT VALUES OF X
      WRITE (NO, 006) (JM, X(JM), JM = 1,P)
006   FORMAT (/, 2X, 3(2HX(, I2, 4H) = , 1PE14.6, 4X))
C
      LAP = 0
      IF (INIT.EQ.0) GO TO 470
      IF (TERM.EQ.1.0) GO TO 480
      IF (LOOP.GE.LOOPY) GO TO 480
      GO TO 1000
C
470   WRITE (NO, 007)
```

```
007   FORMAT (////, 2X, 81HTHE STARTING POINT MUST NOT
      VIOLATE THE CONSTRAINTS. IT APPEARS TO HAVE
      DONE SO.)
480   CONTINUE
490   WRITE (NO, 008)
008   FORMAT (///,2X,29HFINAL DIRECTION VECTOR MATRIX)
      DO 500 J = 1,P
500   WRITE (NO, 009) (J, I, V(J, I), I = 1,P)
009   FORMAT (/,2X,3(2HV(, I2,1H,, I2,4H) = , F10.8,4X))
      WRITE (NO, 011)
011   FORMAT (///, 2X, 16HFINAL STEP SIZES)
      WRITE (NO, 012) (J, E(J), J = 1,P)
012   FORMAT (/, 2X, 3(2HS(,I2,4H) = , 1PE14.6,4X))
C

      END


      FUNCTION F (X, DA, N, NPAR)
C

      DIMENSION X(N), DA(NPAR)
      COMMON KOUNT
C

      X1 = X(1)
      X2 = X(2)
      X12 = X1**2
      X22 = X2**2
C
```

$F = 0.03471 + 0.002703X_1 - 0.017152X_2 - 0.010682\,X_1^2 + 0.01122\,X_2^2 - 0.001788\,X_1\,X_2$

```
C

      KOUNT = KOUNT + 1
C

      RETURN
      END
      FUNCTION CX (X, DA, N, NPAR, K)
```

```
C
      DIMENSION X(N), DA(NPAR)
C
      CX = X(K)
C
      RETURN
      END
      FUNCTION CG (X, DA, N, NPAR, K)
C
      DIMENSION X(N), DA(NPAR)
C
      CG = 0.0
C
      RETURN
      END

      FUNCTION CH (X, DA, N, NPAR, K)
C
      DIMENSION X(N), DA(NPAR)
C
      GO TO (1,2), K
C
    1 CH = 2.0
      GO TO 3
    2 CH = 2.5
C
    3 RETURN
      END
```

A.6 Logic Diagram for the Rosenbrock Program

Rosenbrock[5] has given detailed logic diagram for the Rosenbrock program.

A.7 Table for $f_{k,n}$ Values

Values of $f_{k,n}$ is available from Box and Hunter.[6]

References

1. Khuri AI and Conlon M, Simultaneous optimization of multiple responses represented by polynomial regression function, *Technometrics*, **23**, 363–375, 1981.
2. Box MJ, A new method of constrained optimization and a comparison with other methods, *Computer Journal*, **8**, 42–52, 1965.
3. Kuester JL and Mize JH, *Optimization Techniques with FORTRAN*, McGraw-Hill, New York, 1973.
4. Panda T, (Ed.), *Bioreactors Analysis and Design*, Tata McGraw-Hill Education Pvt. Ltd., New Delhi, India, 2011.
5. Rosenbrock HH, An automatic method for finding the greatest or least value of a function, *Computer Journal*, **3**, 175–184, 1960.
6. Box GEP and Hunter JS, Condensed calculations of evolutionary operation programs, *Techometrics*, **1**, 77–95, 1959.

Further Reading

Kapat A, Synthiesis of extra-cellular chitinase by *Trichoderma harzianum* and characterization of the enzyme, PhD Thesis, Indian Institute of Technology, Madras, India, 1999.
Naidu GSN, Studies on behavior and production of extracellular pectinases from *Aspergillus niger*, PhD Thesis, Indian Institute of Technology, Madras, India, 1999.
Théodore K, Studies on optimization of β-1,3-glucanase production by *Trichoderma harzianum* NCIM 1185, PhD Thesis, Indian Institute of Technology, Madras, India, 1995.

Index